Scientific and Technical Conference on Innovative Mining Technologies (IMTech 2019)

AA006273

IOP Conference Series: Materials Science and Engineering
Volume 545

Szczyrk, Poland
25 - 27 March 2019

Editors:

Marcin Lutynski
Dariusz Prostanski
Tomasz Suponik

ISBN: 978-1-5108-8813-5
ISSN: 1757-8981

Printed from e-media with permission by:

Curran Associates, Inc.
57 Morehouse Lane
Red Hook, NY 12571

Some format issues inherent in the e-media version may also appear in this print version.

Printed by Curran Associates, Inc. (2019)

Additional copies of this publication are available from:

Curran Associates, Inc.
57 Morehouse Lane
Red Hook, NY 12571 USA
Phone: 845-758-0400
Fax: 845-758-2633
Email: curran@proceedings.com
Web: www.proceedings.com

Table of contents

Volume 545

INNOVATIVE MINING TECHNOLOGIES IMTech 2019 Scientific and Technical Conference

25–27 March 2019, Szczyrk, Poland

Accepted papers received: 7 May 2019

Preface

Papers

IOP Conf. Series: Materials Science and Engineering **545** (2019) 011001 doi:10.1088/1757-899X/545/1/011001

INNOVATIVE MINING TECHNOLOGIES

IMTech 2019 Scientific and Technical Conference

March 25th – 27th, 2019

Szczyrk, Poland

The IMTech 2019 Scientific and Technical Conference on INNOVATIVE MINING TECHNOLOGIES was held in Szczyrk, Poland. It gathered over 100 scientists and researchers as well as a big group of specialists representing mining machinery producers and producers of minerals.

The IMTech 2019 Conference was oriented mainly onto innovative and environment friendly technologies and techniques for processing of minerals in the aspect of safety, quality and efficiency.

Five conference sessions contained interdisciplinary presentations on the latest achievements and implementations in the domain of minerals' beneficiation machines and equipment, a beneficiation and an exploitation of minerals, a review of technologies for a low-quality solid fuel gasification, dust control in hard coal processing plants, a remediation and a revitalization of industrial areas, a recovery of metals from e-scrap using gravitational, electrostatic and magnetic separations as well as underground mine workings as potential places for compressed air energy storage and a utilization of methane deposits from liquidated mines in the aspect of environmentally friendly energy technologies.

The conference was an interesting forum enabling an exchange of scientific and professional experience among the participants representing academia and industry. However, the conference series contains only some selected articles, giving an overview of the IMTech 2019 subject-matter.

Editors

Dr. Marcin Lutyński, DSc. Eng. Dr. Dariusz Prostański, DSc. Eng. Dr. Tomasz Suponik, DSc. Eng.

IOP Conf. Series: Materials Science and Engineering **545** (2019) 011002 doi:10.1088/1757-899X/545/1/011002

Peer review statement

All papers published in this volume of *IOP Conference Series: Materials Science and Engineering* have been peer reviewed through processes administered by the proceedings Editors. Reviews were conducted by expert referees to the professional and scientific standards expected of a proceedings journal published by IOP Publishing.

IOP Conf. Series: Materials Science and Engineering **545** (2019) 012001 doi:10.1088/1757-899X/545/1/012001

Coal industry in Czech Republic

V Cablik[1] M Hlavata[2] I Janakova[1] and B Tora[3]

[1]VŠB – Technical University of Ostrava, Institute of Clean Technologies for Extraction and Utilization of Energy Resources, 17. listopadu 15, 708 33 Ostrava – Poruba, Czech Republic
[2]VŠB – Technical University of Ostrava, Department of Environmental Engineering, 17. listopadu 15, 708 33 Ostrava – Poruba, Czech Republic
[3]AGH – University of Science and Technology, Faculty of Mining and Geoengineering Kraków, Poland

vladimir.cablik@vsb.cz

Abstract. Coal reserved in Czech Republic are estimated to be 10 billion tones – bituminous coal about 37%, brown coal about 60% and lignite 3%. Bituminous coal is produced in Northern Moravia – in 2017 production of bituminous coal was 5.5 milion tons. Brown coal is mined in Nord-Wester Bohemia – production of brown coal in 2017 was 38.1 milion tons. Significant quantities of bituminous coal are exported to Slovakia, Austria, Germany and Hungary. In accordance with the National Energy Policy, coal will remain the main source of energy in the country in the future, despite the increased use of nuclear energy and natural gas. The government expects that in 2030 energy from coal will account for 30.5% of energy produced. There are five coal companies in the Czech Republic: OKD, a.s., the only bituminous coal producer and four brown coal mining companies: Severočeské doly a.s., owned by ČEZ, the largest producer of brown coal, Vršanská uhelná a.s., with coal resources until 2055, Severní energetická a.s. with the largest brown coal reserves in the Czech Republic and Sokolovska uhelná a.s., the smallest mining company extracting lignite. OKD operates coal in two mines Důlní závod 1 – (consists of three mines: ČSA Mine, Lazy Mine, Darkov Mine) and Mine Důlní závod 2 (two mines Sever, Jih). The article also presents a pro-ecological solution for the management of waste heaps after coal enrichment - a plant for the enrichment of coal waste from the Heřmanice heap.

1. Coal resources in the Czech Republic

The amount of the geological coal resources, determined by the Czech Geological Survey (Geofond) was estimated at 1519 million tonnes of bituminous coal and 2362 million tonnes of brown coal, of which 181 million tonnes and 871 million tonnes, respectively, were classified as recoverable resources [1,2]. The prospective resources were determined at the level of 6000 million tonnes of bituminous coal and 8820 million tonnes of brown coal [1].

According to the Czech government's energy policy, coal will remain to be the main energy source of the country in the future, despite increasing uses of nuclear energy and natural gas. The government expects that energy generated from coal will reach 30.5% of the total energy produced in 2030 [3,4,5].

IOP Conf. Series: Materials Science and Engineering **545** (2019) 012001 doi:10.1088/1757-899X/545/1/012001

2. Coal deposits in the Czech Republic

2.1. Bituminous coal

The largest bituminous coal deposits are situated in the Upper Silesian Coal Basin. The basin is one of the largest in Europe, with the surface area of 6500 km^2. A larger proportion of that basin is situated within the territory of Poland, with about 1/6 (or 1200 km^2) in the Czech Republic. The Czech area of that basin is called the Ostrava-Karviná Coal Basin. [1]. Figure 1 presents the locations of bituminous coal resources [6].

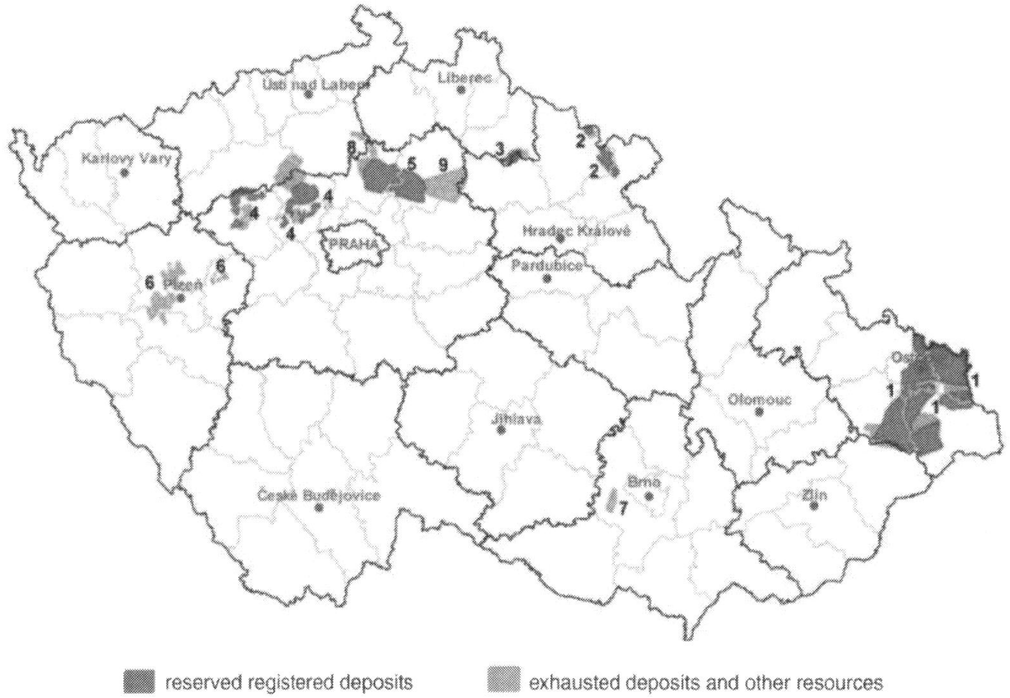

Figure 1. Bituminous coal deposits in Czech [6]: 1. Czech part of the Upper-Silesian Basin, 2. Czech part of the Intra-Sudetic Basin, 3. Krkonoše Mts. Piedmont Basin, 4. Central Bohemian Basins (namely Kladno-Rakovník Basin), 5. Mšeno Part of Mšeno-Roudnice Basin, 6. Plzeň Basin and Radnice Basin, 7. Boskovice Graben, 8. Roudnice Part of Mšeno-Roudnice Basin, 9. Mnichovo Hradiště Basin.

2.2. Brown coal

The Czech Republic has 714 million tonnes of exploitable recoverable brown coal reserves (2016) [6]. In addition to a coal basin in northern Bohemia and another basin near the town of Sokolov, there are coalfields in the south of the country, although they are not economically viable [2]. Production of brown coal totalled 38.6 million tonnes in 2016 [6].

IOP Conf. Series: Materials Science and Engineering **545** (2019) 012001 doi:10.1088/1757-899X/545/1/012001

Figures 2 and 3 present the locations of bituminous coal, brown coal, and lignite resources [6].

Figure 2. Brown coal deposits in Czech [6]: 1. Cheb Basin, 2. Sokolov Basin, 3. North–Bohemian Basin, 4. Czech part of the Zittau (Žitava) Basin.

Figure 3. Lignite deposits in Czech [6]: 1. Vienna Basin, 2. Ceské Budejovice Basin, 3. Czech part of the Zittau (Žitava) Basin.

3. The structure of coal industry in the Czech Republic

Five coal mining corporations operate in the Czech Republic. OKD, a.s. is the only bituminous coal producer [7]. The remaining four brown-coal mining companies are the following: Severočeské Doly a.s., the largest brown coal producer [8], Vršanská uhelná a.s., with the resources available until 2055, Severní energetická a.s., with the largest reserves of brown coal in the Czech Republic [9], and Sokolovska uhelná a.s., the smallest coal mine [10]. The state energy enterprise ČEZ is the largest consumer of coal and an essential provider of electricity in the Czech Republic [2].

4. Organisational structure of the bituminous coal industry

Coal is mined in the Czech Republic only by the OKD a.s. (Ostravsko-karvinské doly) company. In 2016, the bituminous coal output reached 6.7 million tonnes [11], in comparison to 5.5 million tonnes in 2017 [5]. Coal is mined in two plants: Důlní závod 1 (ČSA, Lazy, and Darkov minefields) and Důlní závod 2 (Sever and Jih minefields). Coal mining was ended in Závod Útlum-Jih coal mine (Stařič and Chlebovice minefields) on 31.03.2017 [12].

Figure 4 presents the locations of OKD coal mines. Currently, mining is carried out in the active part of the district, with minefields on the area of 133.65 km^2 [12], totalled 38.6 million tonnes in 2016 [6].

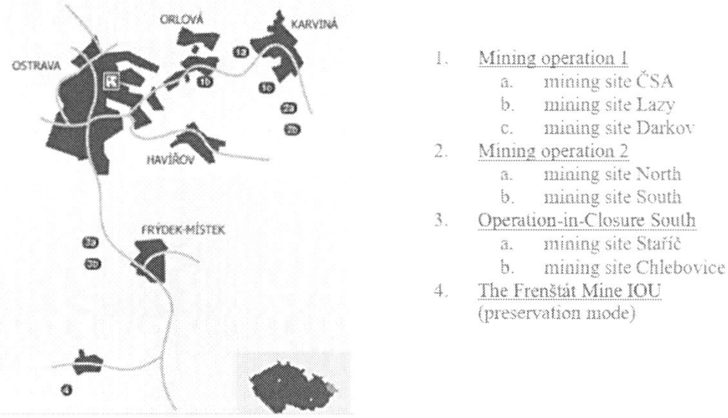

Figure 4. OKD mines location.

The coal layer thickness amounts to 1.5-6.5 m. Mining is conducted by the longwall caving system, with the use of shearer loaders (90.2%) and coal plows (9.8%). Mechanical casing (95.1%) and individual hydraulic casing (4.9%) systems are applied [2]. The OKD company has also been testing the room-and-pillar system as a potential mining method in the Důlní závod 2 coal mine. In each coal mine, coal is agglomerated in processing plants where it is classified as either coking or steam coal, based on specific quality parameters [2]. Coal is agglomerated in the whole grain-size range. Dense-liquid concentrators and dense-medium cyclones are applied as standard agglomeration methods.

5. Brown coal deposits

The Czech Republic has 714 million tonnes of exploitable recoverable brown coal reserves (2016) [6]. In addition to a coal basin in northern Bohemia and another basin near the town of Sokolov, there are coalfields in the south of the country, although they are not economically viable. [2].

6. Mining waste management in bituminous coal mines

Coal extraction has been conducted in the Ostrava-Karviná Coal Basin for about 200 years. Deposition of large quantities of mining waste on heaps in the area of Ostrava is one of the mining consequences. The heaps are subjected to land reclamation processes, with some recovery of energy materials or aggregate production. An example of such waste management projects is found at the Heřmanice storage grounds [13, 14, 15, 16, 17, 18, 19].

IOP Conf. Series: Materials Science and Engineering **545** (2019) 012001 doi:10.1088/1757-899X/545/1/012001

6.1. Heřmanice Heap in Ostrava

The Ostrava heap occupies the total surface area of 100 hectares and it is 250 m high. The heap contains about 21 milion tonnes of mining waste.

The waste processing plant located on the Ostrava-Heřmanice, constructed in 2016-17, is owned by the Ostravská těžební a.s. company [20]. The purpose of that project was to process the waste generated along the bituminous coal production in the Ostrava area in the previous century. The plant is supposed to remove any negative influences of mining waste on the environment. Processing methods allow for obtaining energy coal mixes and aggregates for the building industry. Complete heap land reclamation will be an additional benefit [13, 14, 15, 16, 17, 18, 19].

Owing to the application of knowledge and experience gained from the construction of similar processing plants in other countries, the Ostrava Plant is distinguished by the use of modern technologies and it can be presented as a model of the process line for the heap land reclamation. The process line was designed in the way allowing for the production of construction aggregates, without organic energy components, with the recovery of energy coal mixes in the same process. The company employs about thirty specialists experienced in waste processing, former miners laid off under a restructuring process conducted in the OKD coal mines [16]. It is worth mentioning that several Polish companies participated in the capital investment there: the Institute of Non-Ferrous Metals of Gliwice delivered a floatation machine and the Progress Eco Company delivered screening units. Figure 5 presents a view of the mining waste processing plant.

Figure 5. Coal waste processing plant [16].

In addition, research work was carried out on this heap, which dealt with the flotation of coal waste from the Heřmanice dump. Trčková (2015) tested samples from this site [15]. The flotation reagents Polyethylene Glycol (Foam) and Gyrol (Waste Oil for polishing Si Plates), PP 90 automotive gear oil, and Mogul 15W-40 engine oil were used for testing. The aim was to verify the possibility of flotation from this raw material [15]. Fox (2014) dealt with flotation of waste from the Heřmanice waste rock, where the flotation agents Montanol 800 and Montanol 551 were tested [22]. The aim of the work was to find out if there was sufficient amount of coal in the waste that could be used [21]. This work was followed by Volrábová (2018), whose aim was to optimize coal flotation from the Heřmanice wasteland. The optimization of coal flotation from the Heřmanice waste rock consisted of testing the best concentration of the Montanol 551 [22].

The bachelor thesis, realized at the Department of Environmental Engineering, characterized refurbishment and reclamation projects on the territory of the city of Ostrava, and it was intended to evaluate the course of remediation and reclamation. This thesis contains the legislative conditions necessary for rehabilitation and reclamation, the reclamation division and the process of remediation and reclamation, resulting in an ecologically balanced landscape that is suitable for human life. It also contains specific remediation and reclamation projects that are processed in the city [23]. The work of Janštová (2018) focuses on the Heřmanice dam and the study of the occurrence of selected pollutants. These are mainly heavy metals such as chromium, copper, nickel and zinc [24].

7. Organisational structure of the brown coal industry

Brown coal is mined in the central part of the Czech Northern Brown Coal Basin by two mining companies: Vršanská uhelná a.s. (VUAS) and Severní energetická a.s. (Sev.en), belonging to the Sev.en corporation [7].

The Sev.en manages the largest brown coal deposit in the country, with the ČSA open-pit mine (Figure 6), holding the reserves estimated at 750 million tonnes of good quality brown coal, with the calorific value of up to 17500 kJ/kg. The reserves are adequate for the continuation of brown coal extraction in the next one hundred years. Mining operations will continue within the present deposit boundaries until 2024. 3.6 milion tonnes of brown coal was extracted there in 2015 [25]. The ČSA mining operations were limited by the Government's Resolution on Mining Limits of 1991 (UV No.444/1991), and the validity of the limits was confirmed by the Czech government in October 2015, with a recommendation to revise the decision in 2020 [26].

Besides the ČSA, the Sev.en Group also conducted deep-mine operations in the area of the Centrum Coal Mine (Důl Kohinoor a.s.). However, the business was ended after 128 years of mining on 1.04.2016. That was the last Czech brown-coal deep mine [25]. In 2013, the Sev.en acquired the 800 MW coal-burning Chvaletice Power Plant, which allowed for the establishment of a new integrated Mine and Power Plant organisational structure. After 2015, the Power Plant was upgraded to meet the most rigorous environmental protection requirements. Consequently, its lifetime was extended until 2030 [7].

The Vršanská uhelná a.s. (VUAS), belonging to the Sev.en Energy Group, mines brown coal in the Vršany Coal Mine (Figure 7), located in the central part of the North Bohemian Coal Basin. The local coal resources will allow for the longest period of extraction in the Czech Republic. In 2013, the company signed a fifty-year contract for the delivery of coal to the ČEZ Počerady Power Plant. That long-term contract assures the future of the Vršany Coal Mine and the economic stability of in the north of the Czech Republic [7]. In 2015 r., VUAS produced 6.7 milion tonnes of brown coal, employing 706 workers [27].

The Severočeské doly a.s. (SD) brown coal mine (Figure 8), belonging to the ČEZ a.s. of Chomutov, operates in the north-western part of the Northern Coal Basin and it is located east of the town of Most. SD mines brown coal in two coalfields: Nástup Tušimice and Bílina [28]. The Doly Bílina coal mine produces brown coal with low content of sulphur and delivers it to power plants. The Doly Nástup Tušimice coal mine produces brown steam coal. The SD's annual output amounts to 21 milion tonnes and the company is the largest brown coal producer in the Czech Republic [8].

The Nástup Tušimice brown coal minefields are located between the towns of Chomutov and Kadaň, with one large open-pit mine producing 11,8 milion tonnes of brown coal annually. After coal preparation in the crushing plant, brown coal is sent mainly to the power plants served by ČEZ [29].

The brown coal minefields of Bílina are extracted by one open-pit mine located between the towns of Bílina and Duchcov. Its annual production amounts to 9.9 million tonnes of brown coal. Coal is transported to the agglomeration plant in Ledvice and next to the Ledvice power plant, other industrial companies, and households. In 2017, the SD Group employed 4868 workers. [29] The company pays considerable attention to mitigation of the consequences of mining activities. In the Bílina coal mine, reclamation operations started on 17 hectares and have been completed on 137 hectares, with 32 and 67 hectares, respectively, in the case of the at the Nástup Tušimice coal mine [29].

The coal deposit located in the west of the Czech Republic and in the western section of the brown coal basin close to the Ore Mountain Range (Krušné hory), near the town of Sokolov, is mined by the Sokolovská uhelná coal mine (SU). The company operates one open-pit coal mine, the Jiří. In 2017, the mine's production amounted to 6.9 million tonnes. The brown coal from the Sokolov area is used mainly for electricity and heat generation, although the chemical side products of coal gasification are also important [10].

IOP Conf. Series: Materials Science and Engineering **545** (2019) 012001 doi:10.1088/1757-899X/545/1/012001

The SU company produces electricity and heat in its two plants: the Vřesová IGCC (2 × 200 MWe) and another heat and power plant (5 × 270 MWt), whose total annual production amounts to 3.5 TWh and more than 1800 TJ, respectively. The majority of heat is used by the corporation itself, although some quantities are supplied to the towns of Karlovy Vary, Nejdek, Chodov, Nová Role, Nové Sedlo, and Vintířov. The company also conducts operations for the sake of environmental protection, in particular, land reclamation projects on the lands degraded by open-pit mines, as well as and mining waste processing and disposal. The SU companies employed 1301 workers in 2017 [10].

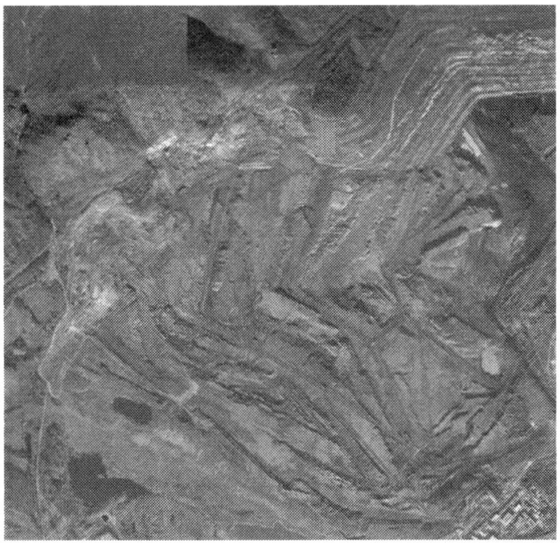

Figure 6. Lom ČSA [30].

Figure 7. Lom Vršany [30].

Figure 8. Severočeské Doly Mine [28].

8. Electricity production and consumption in the Czech Republic

8.1. The Czech Energy Policy

The Czech energy policy is regulated by a number of documents issued by the Czech government:

1) The Czech Energy Policy Conception (SEK); amended in 2015 [3];
2) The National Action Plan for Smart Grids [31];
3) The National Action Plan for Rational Energy Consumption [32];
4) The National Nuclear Energy Development Plan [33];

IOP Conf. Series: Materials Science and Engineering **545** (2019) 012001 doi:10.1088/1757-899X/545/1/012001

The Czech national energy policy is based on the following assumptions: expansion of the existing nuclear power plants; gradual transition from coal- and gas-burning energy generation to that based on renewable energy sources, maintenance of a stable position of coal in the national energy mix (with the reduction of coal share from 45% in 2015 to less than 20% in the decades to come); combined heat and power generation (CHP), stable delivery of coal and other fuels to central-heating installations; increase of energy effectiveness, savings on each form of energy use; reconstruction and development of infrastructure (relating to electricity, gas etc.), integration of decentralised production systems, operating reliability, and energy transit services [3].

The key issues of the Czech energy policy are the following [3]:

1) Diversification of fuel importation (crude oil and natural gas) and expansion of transportation facilities;
2) Acceleration and simplification of administrative procedures for project design and permit issuance, in case of modernisation and construction of new facilities for power-generation infrastructure;
3) Strengthening of international co-operation on energy markets and development of regional markets, in particular in electricity and natural gas trading.

Primary energy consumption amounted to 58.4 Mtoe in the Czech Republic in 2015. Energy generation figures were the following: 39.2% generated from coal (in total: 22.8 Mtoe of which 6.3 Mtoe from bituminous coal and 16.5 Mtoe from brown coal), 16.0% generated from natural gas (9.3 Mtoe), and 21.1% % from crude oil (12.3 Mtoe). Primary energy was supplemented by nuclear energy, in the proportion of 16.4% (9.5 Mtoe), energy generated by sun, water, and wind, in the total proportion of 1.0% (0.6 Mtoe), with biofuels and waste of 6.4% (3.8 Mtoe). [2]

In 2016, about 50.0% of gross total electricity (83.3 TWh) was generated from coal, 29.0% from nuclear fuel and 11.0% from sun, water, and wind [34]. Conventional coal-burning power plants represented the total power of about 10.8 GW. The Czech electricity market was fully opened in 2006 and the gas market in 2007 [2].

The dependence of the Czech market on energy imports has been rather modest; 30.4% of energy demand has been covered by importation [2]. However, import has been structurally unbalanced. The country's dependence on oil and natural-gas imports is estimated at 96-98% [2, 35]. A number of direct and indirect actions were taken to prevent further increase of energy imports, including the following: increase of energy effectiveness, promotion of renewable energy sources on the areas that are the most effective in that respect, with the goal to reach a 13% RES share in the final energy consumption by 2020 [4], as well as an efficient use of local solid fuel resources, mainly brown coal [2]. The directions of import and export of coal are shown in Figure 9.

.

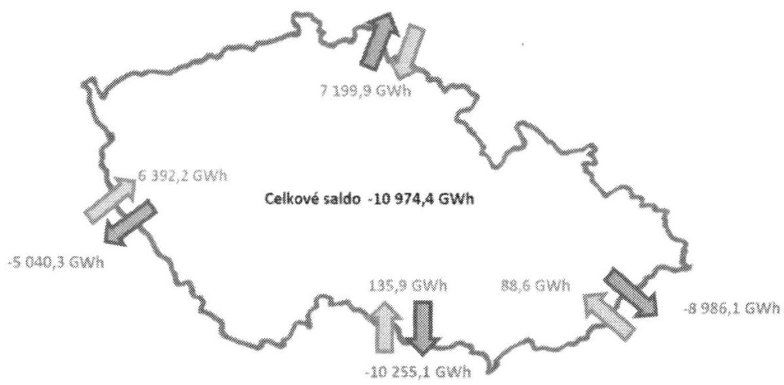

Figure 9. Directions of energy import and export [GWh] [2].

9. Conclusions

The Czech coal mining industry has always played and will continue to play a considerable role in the country's national economy. In 2015, the share of coal in gross electricity generation reached the proportion of 51.4% (gross electricity production refers to the total production from all energy sources) [34], while in 2017 the proportion was only 47% [34]. According to the Czech Energy Policy Conception approved in May 2015, the share of coal in electricity generation should be reduced to 11-21% until 2040 [3]. To ensure a balanced use of coal in that period, the coal-burning power plants were upgraded in the north of the Czech Republic. The 800 MW Tušimice II Power Plant was modernised to reduce CO_2 emission and prolong its operation until 2035 [36]. The lifetime of the Prunéřov II Power Plant was extended by 25 years after upgrades of three power blocks (750 MW), with effective CO_2 emission reduction by 40% [37]. The Ledvice 110 MW power block no. 4 and the new 660 MW power block no. 6 were commissioned in 2017, with their lifetimes planned for 40 years [38].

The Czech Republic is a significant brown-coal producer in Europe (with the extraction of 38.1 million tonnes in 2017, maintained at a stable level since 2010). The extraction of bituminous coal has been dropping each year and it amounted to 5.5 million tonnes in 2017 (ca. 20% less than in 2016) [4].

Acknowledgement
This study was supported by the Institute of clean technologies for mining and utilization of raw materials for energy use, reg. no. LO1406.

References

[1] MCS CR, 2017 - *Mineral commodity summaries of the Czech Republic,* 2017 (Staistical data to 2016). Ministry of the Environement of the Czech Republic/Czech Geological Survey.

[2] Euracoal, 2018 - *EURACOAL. The voice of coal in Europe* 2018, Czech Republic. © European Association for Coal and Lignite, Available fom: https://euracoal.eu/info/country-profiles/czech-republic/

[3] SEK ČR, 2014 - *Státní energetická koncepce České republiky.* Ministerstvo průmyslu a obchodu. Praha 2014. Available fom: https://www.mpo.cz/assets/dokumenty/52841/60959/636207/priloha006.pdf

[4] MPO, 2016 - MPO/ Energetika. *Národní akční plán pro energii z obnovitelných zdrojů. Aktualizace Národního akčního plánu pro OZE schválená vládou v lednu 2016.* Available fom: https://www.mpo.cz/cz/energetika/elektroenergetika/obnovitelne-zdroje/narodni-akcni-plan-pro-obnovitelne-zdroje-energie--169894/

[5] *Energetika v ČR.* 2018 V ČR loni vzrostla těžba hnědého uhlí i výroba koksu. ČTK. 3.3.2018. Available fom: http://oenergetice.cz/rychle-zpravy/v-cr-loni-vzrostla-tezba-hnedeho-uhli-i-vyroba-koksu/

[6] SZ ČR 2017 *Surovinové zdroje České republiky: Nerostné suroviny 2017.* Statistical data. (Česká geologická služba: Praha). , 382 p. ISSN 1801-6693, ISBN 978-80-7075-932-5.

[7] *OKD 2018 The present: here and in the world.* OKD, a. s. https://www.okd.cz/en/coal-mining/the-present-here-and-in-the-world/coal-in-the-czech-republic

[8] Available form: http://www.sdas.cz/spolecnost/profil-spolecnosti/zakladni-informace-o-spolecnosti.aspx,

[9] *Sev.en Energy. Vision and Strategy* 2018 Available form: http://www.7.cz/cz/index.html

[10] *Sokolovská uhelná*, 2018, Available form: https://www.suas.cz/

[11] Konsolidovaná výroční zpráva OKD za období 9.5.2016 - 31.12.2016. OKD a.s. 2017. Available form: https://www.okd.cz/cs/o-nas/vyrocni-zpravy

[12] OKD, 2018 - *OKD. O nás/Kde působí OKD.* OKD, a. s. 2018. Available form: https://www.okd.cz/cs/o-nas/kde-pusobi-okd

[13] Vlášek R 2012 *Využití hořícího odvalu Heřmanice jako netradiční suroviny pro keramické a stavební účely* [online]. Available form: http://hdl.handle.net/10084/95122. (VŠB-TUO:

Ostrava)
[14] Hycnar J 2015 Methods of Increasing the Calorific Value of Fine Coal Waste, *Inzynieria Mineralna – J. of the Pol. Min. Eng. Soc.* **1**(35) p. 33–55.
[15] Trčková I 2015 *Flotace kalů z odvalu Heřmanice* [online]. Available form: http://hdl.handle.net/10084/108026. Diplomová práce. (VŠB-TUO : Ostrava)
[16] Pasiowiec P, Tora B, Wajs J, Bańczyk K., Strączyński L 2018 Sieves as a Basic Element of Processing Processes. 30 Years of Progress Eco Industrial Screens, *Inżynieria Minralna – J. of the Pol. Min. Eng. Soc.* **1**(41), p. 13–34, 2018.
[17] Malíková P, CHromíková J, Čablík V and Thomas J 2015 Flotation Tailings – Intensification of the Dewatering Process by the Aluminum Pre-Hydrolysed Coagulants, *Inżynieria Mineralna – J. of the Pol. Min. Eng. Soc.*, **2**(36), p. 73–78.
[18] Válová, M 2016 *Možnosti využití těžebních odpadů v podmínkách Ostravsko-karvinské aglomerace* [online]. Available form: http://hdl.handle.net/10084/115527. (VŠB-TUO : Ostrava)
[19] KOROLEV I. Coal middlings recycling a route for inceasing the yield of sellable concentrate. *Inżynieria Minralna – J. of the Pol. Min. Eng. Soc.*, **1**(41) p. 159-164, 2018
[20] Ostravská těžební, 2019 Available from: https://www.ostravskatezebni.cz/
[21] Liška R 2014 *Flotace odpadu z odvalu Heřmanice* [online]. Available form: http://hdl.handle.net/10084/103023. (VŠB-TUO : Ostrava)
[22] Volrábová J 2018 *Optimalizace flotace uhlí z odvalu Heřmanice* Available form: http://hdl.handle.net/10084/129929. (VŠB-TUO : Ostrava)
[23] Javorská L 2009 *Sanační a rekultivační práce na území města Ostravy* Available form: http://hdl.handle.net/10084/75868. (VŠB-TUO : Ostrava)
[24] Janštová S 2018 *Studium výskytu vybraných znečišťujících látek na termicky aktivním uhelném odvalu Heřmanice* [online]. Available form: http://hdl.handle.net/10084/130092. (VŠB-TUO : Ostrava)
[25] Konsolidovaná výroční zpráva 2016. Skupina Severní energetická. Available form: http://www.7.cz/cz/spolecnost/downl/Sev.en_VZ_2016.pdf
[26] Usnesení vlády ČR č. 444/1991 ke zprávě o územních ekologických limitech těžby hnědého uhlí a energetiky v Severočeské hnědouhelné pánvi. Úřad vlády České republiky.
[27] Výroční zpráva 2016. Vršanská uhelná a .s. Available form: http://www.7.cz/cz/spolecnost/downl/Vrsanska_uhelna_VZ_2016.pdf
[28] Severočeské doly a.s. 2018 Představení společnosti. Available form: http://www.sdas.cz/prezentace-sdas/start.html
[29] Skupina Severočeské doly a.s., 2017 *Výroční zpráva.* Available form: https://www.sdas.cz/vyrocni-zpravy/SD_VZ_2017.pdf
[30] Sev.en Energy. 2019 Těžba hnědého uhlí. Available form http://www.7.cz/cz/uhli/tezba.html
[31] Národní akční plán pro energii z obnovitelných zdrojů. 2015 Ministerstvo průmyslu a obchodu.. Available form: https://www.mpo.cz/assets/dokumenty/54909/62718/649151/priloha001.pdf
[32] Národní akční plán energetické účinnosti ČR, 2018 Ministerstvo průmyslu a obchodu.. Available form: https://www.databaze-strategie.cz/cz/mpo/strategie/narodni-akcni-plan-energeticke-ucinnosti-cr?typ=tematicky&v=ddc14b4fd2507e907a81dccdd5f9ed12
[33] Národní akční plán rozvoje jaderné energetiky v České republice. 2017 Ministerstvo průmyslu a obchodu/Ministerstvo financí. 2015. Available form: https://www.mpo.cz/assets/dokumenty/54251/61936/640148/priloha001.pdf
[34] Roční zpráva o provozu ES ČR. 2016. Energetický regulační úřad. Oddělení statistiky a sledování kvality ERÚ, Praha 2017. Available form: https://www.eru.cz/documents/10540/462820/Rocni_zprava_provoz_ES_2016.pdf/800e5a09-a58a-4a73-913f-abc30cda42a5
[35] Energy, transportation and environment, 2017 Available form https://ec.europa.eu/eurostat

IOP Conf. Series: Materials Science and Engineering **545** (2019) 012001 doi:10.1088/1757-899X/545/1/012001

[36] CEZ Group. 2019 The Tusimice Power Stations. Available from: https://www.cez.cz/en/power-plants-and-environment/coal-fired-power-plants/cr/tusimice.html

[37] CEZ Group. 2019 *The Prunerov Power Stations*. Available from: https://www.cez.cz/en/power-plants-and-environment/coal-fired-power-plants/cr/prunerov.html

[38] CEZ Group. 2019 *The Ledvice Power Station.* Available from: https://www.cez.cz/en/power-plants-and-environment/coal-fired-power-plants/cr/ledvice.html

IOP Conf. Series: Materials Science and Engineering **545** (2019) 012002 doi:10.1088/1757-899X/545/1/012002

Remediation and revitalisation of industrial areas - a method of promotion

J Biegańska

AGH University of Science and Technology, Faculty of Mining and Geoengineering, Al. Mickiewicza 30, 30-059 Cracow, Poland

biega@ahg.edu.pl

Abstract. Mining and industrial activity causes adverse changes in the environment. The shift from industrial to post-industrial society has shaped the appearance of degraded areas. In a post-industrial society, there is a lower demand for production capabilities and higher for services, consumption and recreation. This study discusses projects involving the remediation and revitalisation of post-industrial post-mining areas.

1. Introduction

The environment is subject to constant changes resulting from human activities. As industrial development progressed, these changes became more and more visible. The 20th century saw a rapid pace of industrial development and was of particular importance in this respect; although the quality of life of the population increased, it had an adverse impact on the environment. Industry and such sectors as metallurgy, mining, power and the cement industry in particular, were the biggest contributors to the degradation of the environment. The most apparent devastation of the environment occurred as a result of the mining industry. It caused deformations of the ground surface, which were either continuous, forming extensive subsidence troughs, or discontinuous – in the form of various types of cave-ins and crevices.

The problem of developing such areas acquired a large-scale importance and 1989 is assumed as the starting period of this phenomenon which was caused, among other things, by the restructuring of the Polish economy enforced by very strict and radical economic and social reforms.

As a result, post-industrial areas and facilities emerging in cities became the subject of various economic activities and were studied by architects, urban planners, geographers, historians of material culture, etc.

Post-industrial areas can be divided [1] into three types:

1. areas that constitute production process components (e.g. excavations, landfills),
2. developed production areas (e.g. warehouses, production halls),
3. areas with surface development (e.g. storage yards, roads and undeveloped right-of-way land of industrial plants).

Table 1 presents the number of industrial areas depending on the number of residents and occupied area [2-4].

Table 1. Post-industrial areas in Polish cities [2-4].

	Number of cities in Poland with post-industrial land (%)							
Number of residents in thousands	100 and more	100-50	50-20	20-10	10-5	5-2	2 and less	average
According to Domański [3]	79.2	73.5	65.4	69.0	44.0	49.5	33.3	58.6
According to Huculak [4]	79.4	69.2	55.3	35.7	40.0	28.1	00.0	46.2
	Total area of post-industrial land in Polish cities (in thousand ha)							
According to Huculak [4]	10.7	4.6	3.5	0.8	2.4	0.2	0.0	22.2

On average, over 50% of Polish cities feature post-industrial areas of more than 20,000 hectares. In addition to post-industrial urban areas, areas abandoned by mining emerged. These are various types of excavations, landslides, landfills, sedimentation basins, etc. At the same time, post-industrial areas, such as areas of military bases and military training grounds as well as farms, formed outside the cities.

Numerous plots of land designated for the above-mentioned activities have become degraded and devastated (these are not only post-industrial areas).

Statistical data [5] prove (table 2) that the number and area of destroyed and degraded areas is decreasing. However, the pace of these changes is low.

Table 2. Devastated and degraded land and remediated and developed land [5].

Specification	2000	2005	2010	2015	2016
	(in hectares)				
Devastated land and degraded land	71.473	64.978	61.161	63.374	64.651
Areas remediated during the year	2.235	1.861	1.222	1.807	1.449
Areas developed during the year	1.222	1.132	581	852	587

Source: data from the Ministry of Agriculture and Rural Development.

In 2017, the degrees of remediation and development of devastated land and degraded land reached the levels of 2.1% and 0.8%, respectively, of the total area of such land which amounted to 62 thousand hectares.

The remediation of post-mining areas is connected with compensating for losses caused by mining activities, but in a substantial number of cases it may also be an opportunity to increase the attractiveness of former mine areas.

2. Key terms

The term "rekultywacja" (remediation, alternatively: reclamation) is defined in the Act on the Protection of Agricultural and Forest Land of 3 February 1995 [6]. Remediation involves creating or restoring utility values in degraded areas – carrying out a series of measures allowing the further use of the land for other purposes, providing it with new social functions.

The remediation process usually consists of three stages:

1. documentation,
2. general (technical) stage and
3. detailed (biological) stage.

Thanks to remediation, post-industrial land may have its previous land use model changed. The most popular types of land use include forests, agricultural and water types. These are the trends most frequently followed in remediation, not only in Poland, but also in other European countries. Less popular, but also present land use types include cultural, economic, natural, residential, educational and recreational.

The term "rewitalizacja" (revitalisation), despite its widespread use, was not introduced into Polish law until late 2015. According to the Act on Revitalisation of 9 October 2015, this term means the process of recovering degraded areas from their existing condition, carried out in a comprehensive manner, through integrated actions for the local community, area and economy, focused on these areas, carried out by the stakeholders of revitalisation in accordance to a certain revitalisation programme [7].

In everyday language, revitalisation (Latin *re* – return + *vita* – life = revival) is understood as activities carried out after the remediation process, carried out in accordance with the local land use plan for a certain area, if such a plan exists.

Revitalisation is an extensive concept, encompassing a number of activities aimed at the repurposing and development of degraded areas. These do not necessarily have to be post-mining areas.

Unlike remediation, revitalisation currently does not constitute a legal obligation of the mining company.

When undertaking revitalisation works, the aim is to sustainably develop previously degraded land and create good conditions for the development of other activities, including, for example, in defunct mining areas. Remediation, on the other hand, is aimed at restoring utility and natural values to land devastated and degraded as a result of various human activities. Remediation is often applied to post-mining areas.

Differences between the processes of remediation and revitalisation also concern the territorial scope – remediation takes place up to the boundary of adverse land transformation, and revitalisation usually concerns vast areas, due to the extant mutual connections, and is not limited only to post-mining areas [8].

The remediation and revitalisation of post-mining areas is a significant economic and technical challenge. Especially when it concerns instances of ground subsidence. Due to limited financial resources, the majority of applied measures are *ad hoc* and involve such actions as levelling the terrain, regulating the water relations and soil restoration. Post-mining waste is often used for land levelling, forming unregulated landfills that can adversely affect the chemistry of surface water and groundwater.

Remediation is the last stage of the activities of a mining business, to which the mining company is obliged by the mining license, but the objectives and directions of remediation should be in line with the methods of subsequent land improvement (revitalisation). In view of the quality of the remediation process, it is desirable for the entrepreneur (if being the owner of the land) to take care of further development and revitalisation.

Revitalisation is considered the highest form of restoring utility functions to areas that used to be exploited by industry.

Areas to be revitalised are described according to the degree of degradation as degraded areas, if their utility value (mainly for agricultural or forest use) has decreased, or devastated areas, if they have completely lost their utility value as a result of industrial activity, among other factors [6].

Depending on the cause of degradation, three types of land transformed by human activity are distinguished, together with their infrastructure.

1. "Brownfield" – a polluted site or part thereof with a potential for development [9]. The term is also used to refer to developed areas for which new functions are sought,
2. "Post-industrial land" - degraded, unused or underutilised area, originally intended for economic activities [10]. K. Gasidło [11] defines post-industrial area as an area where industrial production used to take place and distinguishes three types of such areas: I - the area which was used during the production process; II - area of functions auxiliary to industry, e.g.

administrative and research centres, project offices, water intakes and others; III - industrial impact area, affected for instance by pollution.

3. "Post-mining land" (post-mining areas) – a type of post-industrial land. It is created as a result of the mining of minerals using open-pit, underground or borehole methods. The characteristic signs of the mining process formerly taking place in an area are open-pit and underground excavations, ground subsidence, overburden dumps, mining and processing waste, as well as mine infrastructure facilities. Post-mining areas may form post-mining districts (complexes of post-mining areas), meaning that many facilities of former mines, processing plants as well as processing plants with waste dumps, sedimentation basins and technical and construction infrastructure facilities are concentrated in a small area.

All the above-mentioned areas transformed as a result of human activity should be subject to remediation and redevelopment (revitalisation).

3. Procedures in remediation and revitalisation processes

Remediation is the most important and crucial environmental, social and economic problem. The rational goal of remediation measures is not only to create a stable landscape, consistent with the aesthetics and the surrounding environment, but also to take into account its intended use for future generations. In this sense, the considered remediation (or "reclamation") should take into account the basic objectives which, according to X. Cao [12] include:

- eliminating risks to health and safety (removal of all facilities and structures posing a threat to human health and safety),
- restoring land and water relations (replanting of green areas and stabilisation of residues to reduce mine drainage and water pollution),
- eliminating environmental impact also outside the mine area,
- ensuring a viable, sustainable future for post-mining areas, in terms of environmental, social and economic benefits (development of public areas for recreation, historical purposes, conservation objectives or benefits in the open space or for the construction of public facilities),
- encouraging better use of natural resources.

Mining and land-use are closely linked to a dynamic and integrative process addressed at a range of environmental, production, aesthetic, zoning and economic aspects related to the objectives of remediation planning. This process begins with the opening of a mine and ends when the mine is closed.

Revitalisation is a comprehensive activity aimed at restoring devastated areas for economic use, to the technically possible and economically justifiable extent. The primary problem for undertaking a revitalisation programme is to identify the degraded area and the destructive social, spatial and functional, economic and environmental processes that led to its permanent degradation. The selected indicators should reflect both the nature of the area and the problems that occur there, to the maximum extent possible. The recommended indicators (to designate crisis areas) include[13]:

Social:

- high: unemployment rate, poverty level, crime rate, domestic violence rate,
- low: level of education, number of social organisations (foundations, associations), turnout in elections;

Spatial and functional:

- degradation of the technical condition of buildings, including residential buildings, insufficient equipment in technical and social infrastructure,
- lack of functioning technical solutions enabling effective use of buildings (in terms of energy efficiency and environmental protection),
- lack of access to or low quality of basic services,

IOP Conf. Series: Materials Science and Engineering **545** (2019) 012002 doi:10.1088/1757-899X/545/1/012002

- failure to adapt urban-design solutions to the changing functions of the area,
- low level of communication services,
- shortage or poor quality of public areas;

Economic:

- low level of entrepreneurship and poor condition of enterprises;

Environmental:

- failure to comply with environmental quality standards, the accumulation of waste posing a threat to the life or health of people or the state of the environment.

In revitalisation processes, the following mistakes are usually made [14]:

- generating artificial landscapes,
- inappropriate objectives and assumptions made at the planning stage,
- insufficient communication (dialogue) with stakeholders (mainly with the local community),
- blurring of the cultural landscape,
- disturbance of the landscape with low-value facilities and accidental architectural projects,
- the negative image of the place,
- no attention given to designing the facility's surroundings - no stylistic and architectural cohesion,
- no maintenance and development strategy for the facility/area - operational phase,
- no audit of the benefits of the revitalisation/adaptation carried out – the environmental, social and regional contexts.

A very important and often overlooked element of the adaptation process [15] is the audit of benefits resulting from the transformation carried out in the environmental (the improvement of the environment), social (the positive reception of the project by the local community) and regional (a step forward in the development of the region) contexts.

4. Directions of transformation of post-industrial areas and facilities

Closed-down industrial enterprises often leave abandoned areas and post-industrial facilities behind. Post-industrial zones include many types of areas which have been degraded as a result of various utility functions. These include: landfills, excavations and quarries as well as post-industrial areas, which remain within developed areas [16]. The problem, in this case, concerns large industrial districts and cities, cities not associated with industry, and many small and medium-sized towns.

Activities undertaken in order to solve problems of post-industrial areas and facilities can be divided into three main groups [17]:

1. The protection of the cultural heritage related to industry.
2. The improvement of environmental conditions, landscaping and land use.
3. The improvement (development) of post-industrial land and facilities for economic purposes.

Post-industrial land is usually subject to remediation for the purposes of re-use. These are mostly areas of former mines, gasworks, steelworks and other production plants. Their remediation takes place faster than for instance, the remediation of landfill sites.

The main area of interest of revitalisation programmes include city centre areas, as their use would be fully in line with the strategy of sustainable land management and would allow to protect new undeveloped areas for investments. The possibilities of revitalisation in such areas are consistent with the types of the functions being developed [18]:

- housing and gentrification (the process of improving the social structure of the inhabitants of an area through the inflow of a younger population with a higher social status and the physical improvement of the resources located there),
- shopping centres,
- office facilities,
- hotels and restaurants,

- education and science facilities,
- culture facilities,
- sports and recreation facilities,
- industrial facilities and parks,
- green areas.

The revitalisation of areas previously used by industry has a varying impact on the development of urban tourist space. It depends to a large extent on the role that the city authorities attribute to tourism (especially the tourism related to industrial and post-industrial heritage) in urban policy, and thus on the way in which the new urban spaces acquired by the city are assigned tourist functions. Tourism related to historic technological buildings and post-industrial facilities has achieved unprecedented popularity worldwide [19]; in Poland, it is a relatively new phenomenon, which has not yet been included in the catalogue of tourist products offered to domestic and foreign tourists.

5. Examples of post-industrial land development

5.1. The revitalisation of the Battersea power plant in London

Battersea is a district located on the banks of the Thames in the south-western part of London. The energy industry developed there thanks to the Battersea Power Station. In the 1980s it was included in the national heritage of Great Britain as the largest brick building in Europe and it was decommissioned around that time. It is shown in figure 1.

Figure 1. Battersea Power Station before revitalisation. Photo. Ian Tyas / Getty Images.

Figure 2. The revitalisation of Battersea Power Station in London. Photo: Cushman & Wakefield.

Thanks to the revitalisation, 16 ha of the area will be transformed into a multifunctional building complex (figure 2). The investor Cushman & Wakefield intends to develop and hand over for use 35 ha of modern space with a local park. The London branch of Apple is to be located there.

5.2. The revitalisatioin of the Arcade Providence shopping centre near Boston

The Arcade Providence shopping centre (figure 3) near Boston was one of the oldest of its kind in the U.S. The mall was opened in 1928 and closed in 2008. At that time, a developer became interested in the attractive, historic property and decided to thoroughly modernise the building [20].

As a result of the expansion of online stores in the United States, the interest in shopping centres is declining. The number of facilities closed throughout the country is estimated at about 1200. The developer's innovative project may prove to be an inspiration for the revitalisation of other abandoned shopping malls in the US.

Figure 3.. Arcade shopping centre in the past. Photo by Derek Strahan.

Figure 4.. Shopping mall converted into cheap flats. Photo by arcadeprovidence.com.

The Arcade building was rebuilt according to the revitalisation design of Northeast Collaborative Architects. Shops on the first floor are replaced with 48 small flats with an area of 21 to 42 m^2 (figure 4).

The revitalisation project of the building combined a recreation and entertainment area located on the ground floor of the building with micro apartments on the first and second floor of the former shopping centre (figure 5).

Figure 5.. The recreational and commercial zone on the ground floor of the building. Photo by Christine Francis.

6. An example method of promoting post-industrial areas

In view of the fact that tourism has been increasing in recent years, it seems possible to attract the public to remediated areas by applying various incentives. Some of the possible options for promoting both developed and undeveloped post-industrial areas may include a combination of geotourism and geocaching. [21]. A box (a "cache") hidden by other participants of the game, is a container protected against moisture, containing the visitors' logbook. The people who prepare these boxes make sure that their location is attractive for those who look for them. A cache is often hidden in places with an interesting history, which the hiding person describes in the profile of the mailbox on the website.

Post-industrial areas can be promoted in a similar way by placing a cache in their vicinity, which encourages people to learn about the history of a certain place. Sometimes it is the only way to reach the general public and draw attention (interests) to places of historical significance.

A telling example of generating publicity around a forgotten post-industrial area is the history of a brick factory, which was promoted by J. Rogóż [22] using the method described above.

6.1. Tarnów - brick factories

At the beginning of the 20th century there were five brick factories operating in Tarnów:

1. Rudy – it is estimated that it was established before 1865. The plant ceased its activity before the Second World War. The excavations left overgrown wastelands, remainders of buildings in the form of: outbuildings, administrative and storage facilities and drying rooms.
2. Konstancja – one of the larger and more popular brickyards operating in Tarnów in the 20th century. It was closed in September 1987. The historic architecture was demolished. Only a few devastated buildings and a chimney remain.
3. Kantoria – established in 1902-1906. It produced bricks and tiles for many years, for various owners. Production was discontinued in 1980. A chimney and an unprotected and unused excavated tank remained after the plant.
4. Tarnowianka - located in the vicinity of the Konstancja plant. Construction works were completed in 1908. In the 1950s the brickyard was modernised and operated until the 1990s, when the clay deposits were exhausted. At present, no remains of the buildings have been preserved; new halls have been built in their place, and the town's southern ring road runs through the former excavation areas.
5. Mieszczanka –probably established around 1912. The name derives from the fact that the brickyard was a municipal factory. In 1951, Mieszczanka became a part of Krakowskie Zakłady Ceramiki Budowlanej. It was modernised and operated continuously, despite the exhaustion of raw materials in its excavations, until 1989, when the operation was discontinued.

In the area of the former cement industry areas there remain a post-mining pit (figure 6), overgrown wastelands, devastated buildings (figure 7) and a chimney (figure 8).

Figure 6.. Kantoria post-mining pit. Photo by P. Chwał.

Figure 7.. Other buildings – Kantoria. Fig. S. Siekierski.

Plans for redeveloping the water reservoir of the former Kantoria brickyard assume the construction of an open amphitheatre for outdoor events (figure 9), arranging a beach and providing access to the reservoir for water sports enthusiasts. It is also planned to build an amphitheatre with a floating stage and a promenade along its banks. The project [23] also provides for the creation of an educational path, an outdoor gym and swimming pools and wading pools.

IOP Conf. Series: Materials Science and Engineering **545** (2019) 012002 doi:10.1088/1757-899X/545/1/012002

Figure 8. The chimney of the Konstancja brick factory. Photo by J. Rogóż.

Figure 9. The planned appearance of the surroundings of Kantoria after the reconstruction. Photo: visualisation/ archive of the Tarnów City Hall.

6.2. Setting up caches

In designing caches meant to preserve the memory of the former brickyards, a convenient location was chosen. The locations of caches  are indicated on the map of Tarnów (figure 10). It was decided that most caches would be located close to trees. The caches had the form of "breakfast boxes", i.e. tightly closed plastic containers.

In order to make the caches more attractive, the first finders could obtain pins with photographs of the brickyard and their most characteristic elements, i.e. chimneys, from each cache. During the inspection of the area, it turned out that one of the brickyards already has its own cache.

Figure 10. Location of the author's cache in the field [24].

6.3. Feedback from caches

Table 3 contains data on geocaches with their numbers in the Geocaching portal.

Below are examples of comments made by Geocaching users addressed to the author's caches (original spelling).

Kantoria:

Commentary by Euphoria_: "Luckily, the author is online and called with a hint. Kantoria. A beautiful story, beautiful remains Thanks for the cache and {FTF}" [FTF- first to find- author's note]

Mieszczanka:

Commentary by Euphoria_: "Fast bus ride from Kantoria to Mieszczanka. A nicely prepared cache. Thank you for the story of Mieszczanka and for the third, great pin."

Table 3. Statistics of caches visits [22].

Cache name	Geocaching number	Number of people searching (as at 4 July 2018)
Rudy	GC7QJ57	5
Mieszczanka	GC7QBEC	8
Tarnowianka	GC7QBE6	5
Kantoria	GC7qBDQ	8

7. Summary

The presented examples regarding post-industrial areas show that attractive revitalisation can be achieved in a non-standard way. The potential of the land in question is used for this purpose, both in terms of its utility and economy. It receives new functions thanks to the revitalisation process. Both in Poland and across the world, industrial heritage is becoming a showcase of certain regions and a tourist attraction.

Post-industrial areas require a comprehensive approach during the revitalisation process. In many cases, their revitalisation reduces the negative impact caused by the disappearance of jobs, minimising social frustration.

It is a new approach to promote revitalised areas. This is a kind of response to the emerging trend of industrial heritage tourism.

By engaging in geocaching, tourists have the opportunity to reach places and facilities that are not mentioned in traditional tourist guides or included in the offers of travel agencies.

Due to the fact that the geocaching network covers almost the whole of Poland, the tourist has the opportunity to start and finish a sightseeing tour at almost any place and time, choosing the most interesting sites on the way.

Geocaching is currently a niche phenomenon, which, however, has the potential to become more widely known. Thanks to this, the previously hidden industrial heritage, with only minor investments, could become a real tourist attraction. Geocaching is a young but very interesting form of tourism. Users are attracted by the need for conspiracy and the aura of mystery that accompanies the search. Geocachers gather in social media and on the geocaching.com website. New applications are created, participants in the game are awarded badges, regular events are held to familiarise the participants with a given topic (city/region). People interested in geocaching are usually highly-educated individuals who have been involved in this activity for years [25]. For some people it is a passion to which they dedicate most of their free time, and consider the presence of geocaches in an area the decisive factor when choosing their holiday destination.

References

[1] Gasidło K 1998 *Zeszyty Naukowe Seria Architektura Wydawnictwo Politechniki Śląskiej* **37**
[2] Gasidło K 2018 *Człowiek i Środowisko* **41(3)** pp 5-29
[3] Domański B 2000 Restrukturyzacja terenów poprzemysłowych w miastach Ziobrowski Z et al. (ed), *Rewitalizacja. Rehabilitacja. Restrukturyzacja. Odnowa miast, Instytut Gospodarki Przestrzennej i Komunalnej*
[4] Huculak M 2009 Rewitalizacja terenów poprzemysłowych Polskie doświadczenia i perspektywy Jarczewski W (ed) *Przestrzenne aspekty rewitalizacji Instytut Rozwoju Miast*
[5] Główny Urząd Statystyczny 2017 *Ochrona środowiska* Warszawa
[6] Journal of Lows (1995, no 16 item 78)
[7] Journal of Lows (2015, item 1777)

[8] Kasztelewicz Z 2012 Ekonomiczna rekultywacja i rewitalizacja terenów pogórniczych *Kopaliny* **1** pp 16-20

[9] Glapa W 2004 O potrzebie aktualizacji przepisów dotyczących rekultywacji gruntów i terenów. 10 Konferencja GSG'04 Szkody, odszkodowania i zabezpieczenia roszczeń na terenach górniczych Polanica Zdrój

[10] Program rządowy dla terenów poprzemysłowych przyjęty przez Radę Ministrów w dniu 27 kwietnia 2004 *Ministerstwo Środowiska* Warszawa

[11] Gasidło K 1998 Problemy przekształceń terenów poprzemysłowych *Zeszyty Naukowe Wydawnictwo Politechniki Śląskiej* **1408**

[12] Cao X 2007 Regulating Mine Land Reclamation in Developing Countries: The Case of China *Land Use Policy* **24** pp 472-483

[13] Instrukcja dotycząca przygotowania projektów rewitalizacyjnych w ramach Regionalnego Programu Operacyjnego dla Województwa Mazowieckiego na lata 2014-2020 oraz preferencji dla projektów mających na celu przywrócenia ładu przestrzennego Version 1.1 Warszawa 2016 *Urząd Marszałkowski Województwa Mazowieckiego w Warszawie*

[14] Kobylańska M and Gawor Ł 2017 Problematyka przeobrażeń przestrzennych w procesach rewitalizacji terenów poprzemysłowych *Prace Komisji Geografii Przemysłu Polskiego Towarzystwa Geograficznego* **31**(1) pp 68-80

[15] Pietrzyk-Sokulska E 2005 Kryteria i kierunki adaptacji terenów po eksploatacji surowców skalnych – studium dla wybranych obszarów Polski *Wydawnictwo Instytutu Gospodarki Surowcami Mineralnymi i Energią PAN*

[16] Lorek A 2008 Rewitalizacja terenów poprzemysłowych i pogórniczych w regionie śląskim (ed) D. Kotlorz, *Wyd. Akademii Ekonomicznej*

[17] Gasidło K 2008 Przekształcenia terenów poprzemysłowych – efekty i perspektywy badań i działań *Problemy Ekologii* **12**(2) pp 76-80

[18] Bryxa M and Jadach-Sepioło A 2009 Instytut Rozwoju Miast Rewitalizacja miast w Niemczech *Wyd. Instytut Rozwoju Miast* **3**

[19] Jędrysiak T 2011 Turystyka kulturowa w obiektach poprzemysłowych – zagadnienia ogólne *Turystyka Kulturowa* **6** www.turystykakulturowa.org

[20] Zamieszkaj w centrum handlowym – sprytny projekt rewitalizacji z USA https://www.morizon.pl/blog/zamieszkaj-w-centrum-handlowym/, dostęp 23.08.2016

[21] Biegańska J 2018 Promowanie zrekultywowanych terenów pogórniczych. Innowacyjne i przyjazne dla środowiska techniki i technologie przeróbki surowców mineralnych Bezpieczeństwo – Jakość – Efektywność *KOMEKO* pp 88–89

[22] Rogóż J 2018 Geocaching – nowe możliwości propagowania terenów poprzemysłowych. Praca dyplomowa magisterska, *AGH Wydział Górnictwa i Geoinżynierii*

[23] Chwał P 2018 Tarnów zyska urokliwe miejsce do wypoczynku, rekreacji i koncertów, *Gazeta Krakowska* https://gazetakrakowska.pl

[24] Zabawa 2018 https://www.geocaching.com and https://www.geocaching.com/map/ dostęp 02.04.2018

[25] Samołyk M 2017 Tajna turystyka w XXI wieku – dekada geocachingu w Polsce, *Journal of Education Health and Sport* **7**(7) pp 1084-1094

IOP Conf. Series: Materials Science and Engineering **545** (2019) 012003 doi:10.1088/1757-899X/545/1/012003

Coal Images Database and Its Application for Selected Coal Grains Properties Analysis

S Iwaszenko and L Róg

Central Mining Institute, Plac Gwarków 1, 40-166 Katowice, Poland

s.iwaszenko@gig.eu

Abstract. The raw materials processing methods play important role in excavation industry. The area is nowadays a target for novel, innovative technology application. One of them is an image processing as an instrument for determination of different material properties. The paper proposes a tool – database system – dedicated for supporting images processing and computer vision methods application in the mining industry. The idea and implementation of the system is presented. The database was supplied with a set of coal images. The images were taken at a special stand, with the calibrated camera. The usefulness of system was tested by application of a pipeline of image processing algorithms. The goal was to estimate the grain size composition of coal and rock matter presented on images. Two algorithms for edges detection were tested: Sobel filter and Statistical Dominance Algorithm. After edges detection, a set of morphological operations were applied of determination of good starting markers for the watershed algorithm. The result showed that both algorithms tend to produce over segmented images. However, SDA performed noticeably better. The experiments proved the usefulness of developed database system dedicated to image processing.

1. Introduction

Coal and in general rock processing methods are one of the crucial steps in the raw material industry. The quality and classification of the solid fuel are strictly connected with its physicochemical properties. The determination of coal's properties usually involves a significant amount of human labor. Therefore, application of methods offered by information and communication technologies can help in raising the effectiveness of analysis and improve their results. The image processing and analysis is a discipline which application in mining industry becomes more and more important. The application of image processing in Run of Mine (ROM) and aggregate analysis is growing since the 1990s. The image processing was used for grain size distribution determination [1, 2], ore classification [3], coal gangue identification [4] and so on. Despite the machine vision method used, the effective and efficient way of information storage and retrieval is necessary. There is also need for preparation of rock grains image database, which can be used for development and testing of dedicated computer vision methods. Unfortunately, the images' sets used in an application of computer vision to mineral raw materials processing methods are hardly reused. One of the rear examples of such reuse is given by [5]. The reported research uses the same images and images' preparation procedure as reported in [3]. In much other application of image processing techniques, the usage of common picture databases is usual approach. As an example of available databases the IMAGENET repository [6] or databases dedicated to emotion or face recognition. However, it seems that there is lack of such

IOP Conf. Series: Materials Science and Engineering **545** (2019) 012003 doi:10.1088/1757-899X/545/1/012003

a database in case of aggregate or crushed rocks images. The authors developed a database system for coal and coal accompanying rock images. Though the idea for application of database technology is not new – an excellent attempt was described in [7] – there is no images database directed to solid fuels. Authors believe that proposed solutions extend the idea presented in [7] and directs it towards coal mining industry. The proposition of database system is presented. The system was implemented and a feed with coal samples images. The application of the developed system to the aggregate grain sizes determination by batch images processing was presented. The goal of the work was to create a dedicated but elastic tool helping in image processing techniques application in mine industry.

2. Material and methods

2.1. General assumptions and ideas

The computer vision methods application in any discipline is involved with processing a certain number of images. The concept of application of database technology to support management of the images' set given in [7] suggests following approach to the problem:

- Database systems are targeted in processing large amounts of data, therefore they are a natural choice as a platform for bulk machine vision application.
- The processing effort should be moved from client computers to database servers thus improving mobility and lowering requirements for end user's equipment.
- Machine vision algorithms should be implemented in form of User Defined Functions, which repository should be stored at the database server.
- There should be possible to write images formed as a result of image processing method application back to the database.
- The results of machine vision application should be stored as a set of parameters in the database, forming the knowledge base.
- All data stored in the database system should be available for end users and for further processing.

The authors present the results obtained by application of proposed idea – the research on retrieval of information about grain sizes in copper concentrate as well as segmentation of grains in thin sections. However, there was no information about database system architecture nor information about its implementation. In particular, there was no clear statement if the same database system (though selecting an appropriate set of images) was used in both types of research. The proposed work addresses this area of database system dedicated to computer vision applications. Usefulness of such system is recognized in two areas:

- The system should be capable of batch processing of the big amount of visual information stored in the database by use of processing workflow (batch mode).
- The system should deliver the functionality for performing on demand, individual processing of selected images by use of desktop or web application (interactive mode).

The results of analysis should be stored in the same database. The presented areas are not separated. In fact, the interactive mode is a tool for prototyping and establishing the appropriate sequence of operations, which lead to achieving the desired result. Once the image processing and analyzing algorithm are established, the workflow for batch mode can be prepared. Then, the process is performed on the server side, for possibly a long time. During the process, the selected bunch of images are transformed and analyzed automatically. The results of operations are stored in the database for further investigations and usage. The schematic idea of operation is presented in figure 1.

The processing workflow, stored in the database system, can be easily reused. Not only does it allow efficient processing of any new images stored in the database, but also form the basis for further development.

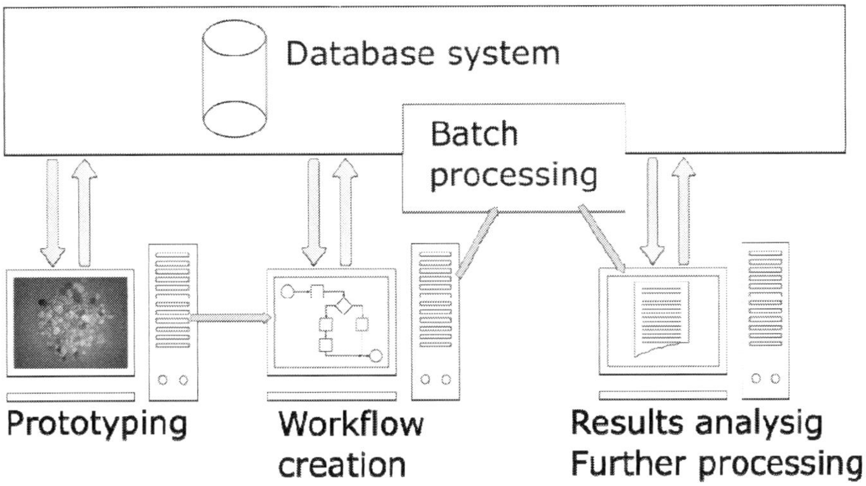

Figure 1. Schema of database usage for images processing and analysis.

2.2. Database system architecture and implementation

The proposed database system was divided into few modules, responsible for important aspects of functionality. The modules can be characterized as follows:

The data repository, responsible for storing all data and information processed and shared by the system. The data repository stores both: images and all kind of information describing the images. This includes the results of the application of image processing algorithms to images as well as results of the application of analytical tools. The data repository logs all operations and workflows which were used to process the images.

Image processing and analysis routines repository stores software components in form of object-wrapped processing procedures. The solutions were preferred to suggested in [7] User Defined Functions. The objects implement a unified interface, thus facilitating the composition of complex processing sequences. The objects can be relatively easily exchanged, according to Liskov's substitution principle [8].

The workflow engine is responsible for the execution of the sequence of image processing or analysis operations. The sequence (workflow) itself is managed by the workflow engine. Components of the sequence are taken from image processing and analysis routines repository. Workflows can have branches. It is suggested that workflow engine is based on one of available workflow generic solutions, e.g. Windows Workflow Foundations (WWF).

Batch processing manager is responsible for orchestrating of batch processing. It should start at scheduled time image processing tasks. Each task executes a workflow for a set of selected images given as parameters.

Application Programming Interface (API) forms an interface between system users and system. API provides a set of functions with well-defined parameters, which gives an access to database system functionality. Client applications, operated by end-users, interact with the system through this component.

The overall system architecture was presented in figure 2.

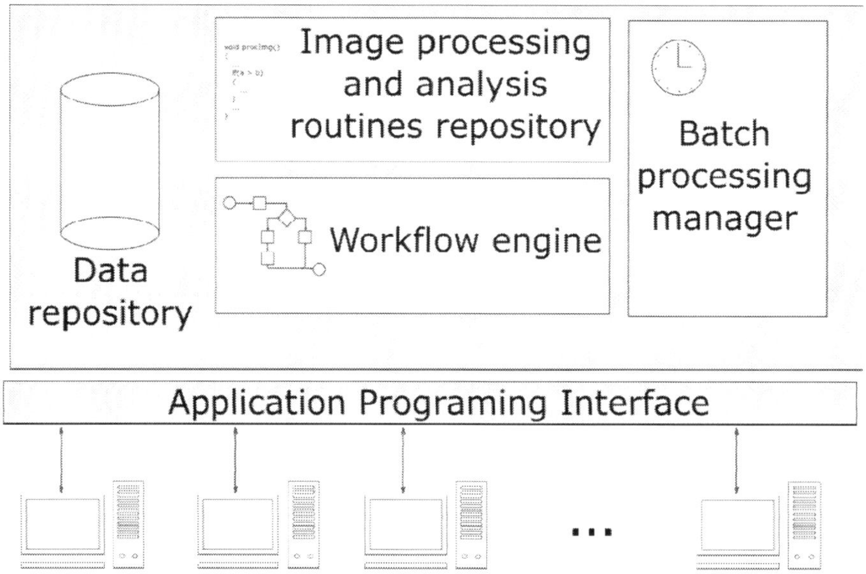

Figure 2. Database system for images processing architecture.

The image processing database system was implemented using Microsoft tools and technologies. The Microsoft SQL Server 2017 was used as a relational database management system. Workflow engine was developed with Windows Workflow Foundation and SQL Server agent was used for batch processing management. The image processing and analysis routines were organized as a filesystem stored dynamically linked libraries, using MEF (Managed Extensibility Framework).

2.3. Coal images acquisition
The coal images were taken at Central Mining Institute in the Department of Solid Fuels Quality Assessment. The images were taken using Nikon D80 digital camera. For images acquisition, the dedicated stand was prepared. All samples taken for analysis were photographed at different camera focal length. Pictures were taken in the ambient daylight conditions. The images have a resolution of 3872x2592 pixels with 8-bit RGB color space. The camera was calibrated using chessboard template [9]. The images were stored in the database along with the sample code for further reference. There were was also a linear measure placed on each photograph. The images were saved in a lossless format.

3. Results and discussion

3.1. Coal images database
During several images, acquisition sessions over 500 of images were taken and stored in the database system. The photographed samples consisted of enriched and sieved coal, coal and barren rock with mixed granularity and coal originating wastes. There were also few pictures of crushed Run-Of-Mine (ROM). Because of imaging scene character, the automatic exposure parameters selection could not be directly used. For each sample, a basic information of its origin, as well as description, were stored. For the majority of samples laboratory, fuel analysis results are also available. The typical photograph gained is presented in figure 3.

Figure 3. Typical example of images stored in the database system.

3.2. Images processing test

The developed solution was used as a tool for investigations on selected grain sizes detection algorithms effectiveness. The overall procedure used was schematically presented in figure 4.

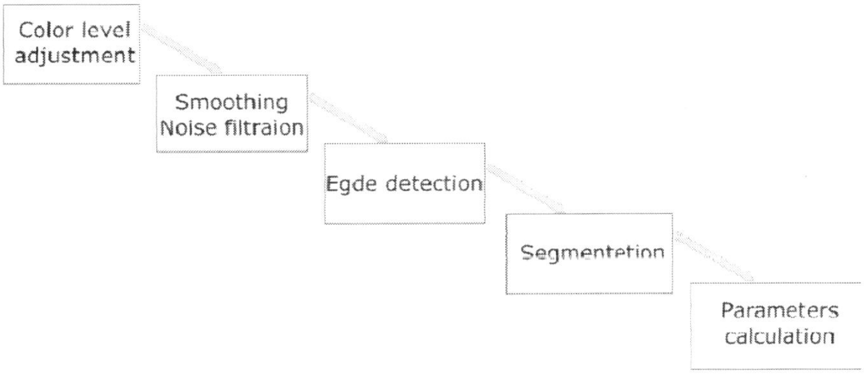

Figure 4. Image processing sequence.

The selected images stored in the database were cropped to ensure that there are no other objects than coal or rocks visible. The resulting pictures had the sizes of 3872x2096. Then the images were transformed to HSV color space and only the V (Value) channel, representing image brightness was taken for further processing. The median filter with square structuring element and sizes ranging from 3x3 to 11x11 were applied to each monochromatic picture. The resulting images were the basis for edges identification. Two algorithms were used for this purpose: the Sobel filter and Statistical Dominance Algorithm. The Sobel filter is one of the commonly known methods of edges detection. It exploits the observation that edges are usually the places of abrupt changes in pixels brightness. Therefore, the length of gradient vector can be treated as an edge indicator [10].

$$|\nabla I(x,y)| = \sqrt{\left(\frac{\partial I(x,y)}{\partial x}\right)^2 + \left(\frac{\partial I(x,y)}{\partial y}\right)^2} \quad (1)$$

As long as gradient magnitude is greater than the selected threshold, the points are treated as belonging to the edge. The method, though simple, is not resistant to the artifacts. There is also not always clear which value of thresholding should be used.

SDA is one of the recently proposed algorithms for edges and structures detection. The algorithm focuses on differences in grey level between the selected point and its neighborhood. The number of points exceeding the selected point value is the result value of the operation. There is also possible to take into account only this points from the vicinity, which values are greater of selected point value increased by given threshold. The formal description of the algorithm is presented in the equation below [11].

$$\bigwedge_{p(x,y)\in I} \sum_{p_b(x_b,y_b)\in B(x,y)} \begin{cases} p'(x,y) := p'(x,y) + 1, p_b(x_b,y_b) \geq p(x,y) + t \\ p'(x,y) := p'(x,y), otherwise \end{cases} \quad (2)$$

where $p(x, y)$ is a pixel from the source image, $p_b(x_b, y_b)$ and is a pixel belonging to neighborhood $B(x, y)$ of point $p(x, y)$. The $p'(x, y)$ is resulting pixel and t stands for the grey level threshold. The resulting image is normalized and then, as in the case of Sobel filter, a threshold is applied. The threshold level was selected individually for each image. For that purpose selected, representative set of images were processed manually. The edges of aggregate grains were drawn by trained personnel. After that, the average edges` area ratio was calculated yielding 9.3%. As the Sobel/SDA detected edges are fragmented it was assumed, that about half of the edges are missing. Therefore, the threshold level was chosen to ensure that no more than 4.5% of all pixels in given image will be selected. The thresholded images are processed by dilation algorithm to remove tiny elements. Again, the process was continued until at least 10% of initial thresholded areas were eroded. Than a watershed algorithms are applied, using remaining areas as seed. For the resulting segmentation, a histogram of grain sizes for each picture is calculated. The histograms obtained by Sobel filter edge detection method as well as SDA algorithm were compared.

3.3. Tests results

The exemplary result of proposed processing for the same image once using the Sobel filter and once SDA algorithm is presented in figure 5 and figure 6 respectively. The histograms for presented image are presented in the figure 7 and figure 8.

Figure 5. Results of coal image processing for Sobel filter (selected image). a – original image, b – image processed by 11x11 median filter, c – the normalized output of Sobel filter, d – watershed segmentation result.

Figure 6. Results of coal image processing for SDA filter (selected image). a – original image, b – image processed by 11x11 median filter, c – the normalized output of Sobel filter, d – watershed segmentation result.

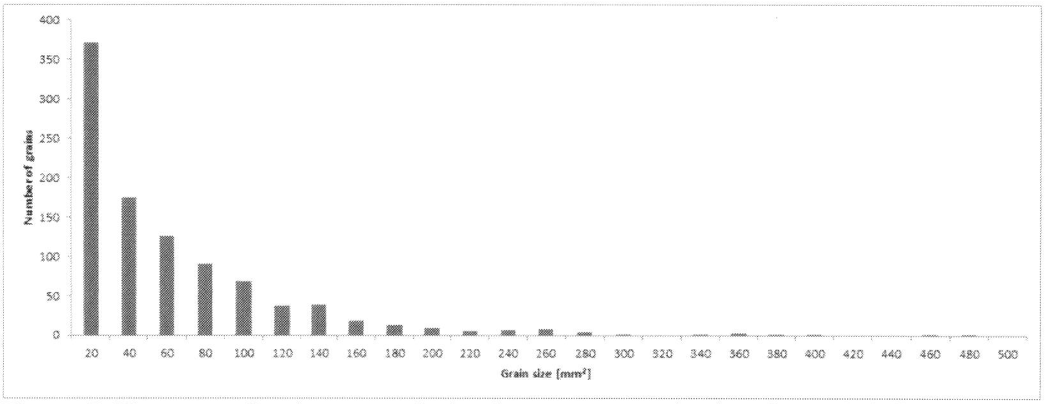

Figure 7. Histogram of grains areas calculated upon segmentation based on Sobel filtered image.

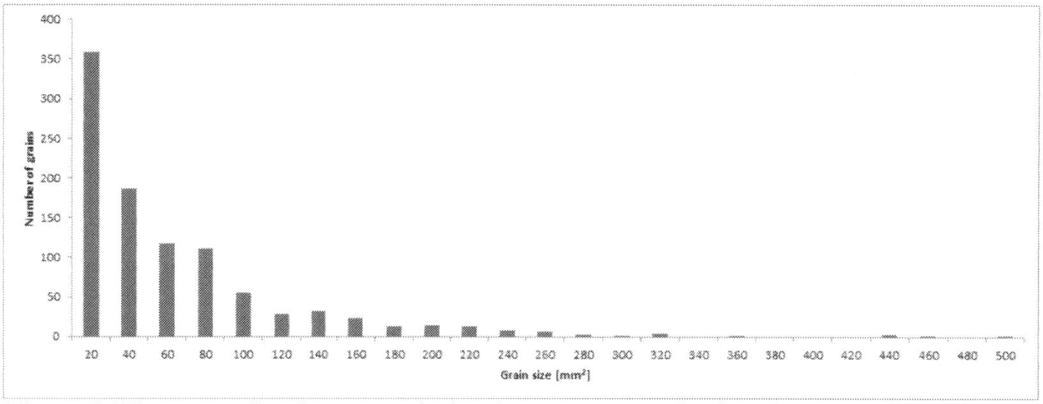

Figure 8. Histogram of grains areas calculated upon segmentation based on SDA filtered image.

It can be observed, that despite the method used the watershed algorithms tend to over segment the image. The resulting distribution is therefore shifted towards lowest values in comparison to expert determined segmentation. The SDA algorithm seemed to perform slightly better, resulting in lesser over-segmentation. Despite that, SDA resulted in segmenting grains of extremely large areas. It suggests a modification to the algorithm used: after an object exceeding threshold size is detected, it is processed again for better segmentation. Nevertheless, the testing showed that developed database of images allows for relatively easy bulk processing of a big number of images. Though the processing times were relatively large, the server-oriented processing gives the possibility to use server resources for calculation. There is also possible to schedule the processing to off hours and collect the results in suitable time.

4. Conclusions

The coal images database was developed for coal images storage and processing. The database was supplied with coal and barren rock images taken during laboratory samples acquisition. The images were stored with information allowing their matching with measured coal properties. There are above 700 pictures of coal with different granularity and properties stored in the database. A set of image processing operations were performed showing the usefulness of the designed database. The images were filtered for noise removal by the median filter, then the grain edges were detected using Sobel or SDA filter. After that, the grains were segmented using the watershed algorithm. Though obtained results showed significant over-segmentation, they proved that the developed system can be efficiently

IOP Conf. Series: Materials Science and Engineering **545** (2019) 012003 doi:10.1088/1757-899X/545/1/012003

used for batch images processing. The further development of the system is planned as well as research aimed at obtained precise image segmentation methods. The further research will also consider other features of coal grains as color (in various color spaces), texture, grain shape properties and so on. It is expected that enhanced features set will allow not only shape and size of rock grains determination but also coal type recognition.

Acknowledgments
The works presented in the paper have been supported by the statutory activity of the Central Mining Institute: Application of image recognition and machine learning methods for Run-Of-Mine and waste mineral matter characterization – No. GIG: 11010117-144.

References
[1] Maerz N 1998 Aggregate sizing and shape determination using digital image processing *Center For Aggregates Research (ICAR) Sixth Annual Symposium Proceedings* pp 195–203
[2] Persson A L 1998 *Engineering Geology* 50 177 – 186 ISSN 0013-7952 URL http://www.sciencedirect.com/science/article/pii/S001379529800009X
[3] Tessier J, Duchesne C and Bartolacci G 2007 *Minerals Engineering* 20 1129 – 1144 ISSN 0892-6875 URL http://www.sciencedirect.com/science/article/pii/S0892687507001161
[4] Wang R and Liang Z 2011 Automatic separation system of coal gangue based on dsp and digital image processing *2011Symposium on Photonics and Optoelectronics (SOPO)* (IEEE) pp 1–3
[5] Perez C A, Estévez P A, Vera P A, Castillo L E, Aravena C M, Schulz D A and Medina L E 2011 *International Journal of Mineral Processing* 101 28 – 36 ISSN 0301-7516 URL http://www.sciencedirect.com/science/article/pii/S0301751611001037
[6] 2019 URL http://www.image-net.org/
[7] Mlynarczuk M, Ladniak M and Piorkowski A 2014 *Physicochemical Problems of Mineral Processing* 50(**2**) 563–573
[8] Liskov B and Wing J M 1993 Family values: A behavioral notion of subtyping Tech. rep. Carnegie-Mellon Univ Pittsburgh PA Dept of Computer Science
[9] Zhang Z 1999 Flexible camera calibration by viewing a plane from unknown orientations *Computer Vision, 1999. The Proceedings of the Seventh IEEE International Conference on* **1** (IEEE) pp 666–673
[10] Szeliski R 2010 *Computer vision: algorithms and applications* (Springer Science & Business Media)
[11] Piorkowski A 2016 *A statistical dominance algorithm for edge detection and segmentation of medical images* (Springer) pp 3–14

IOP Conf. Series: Materials Science and Engineering **545** (2019) 012004 doi:10.1088/1757-899X/545/1/012004

The influence of the type of fly ash on selected physical properties of their water mixtures (multiphase mixtures)

P Pierzyna

Silesian University of Technology, Faculty of Mining and Geology, Akademicka 2, 44-100 Gliwice, Poland

piotr.pierzyna@polsl.pl

Abstract. Waste from energy or mining has been widely used in civil engineering for many years, ie in road engineering, construction, land reclamation as well as in mining. The most frequently performed technologies in underground mining with fine-fraction waste are the sealing of infarcted grouting of cavings and liquidation. In these technologies, transport and migration properties play a primary role. They depend on the type of fly ash. At present, in Polish hard coal mines located in Upper Silesia, most commonly used technologies are ashes with code 10 01 02 from conventional boilers, and as second ashes from code 10 02 82 from fluidized bed boilers. The article presents the results of research on selected physical properties of ash-water mixtures such as: water-ash index (W/FA), leveling, density and volume of excess water. Furthermore, for the tested ashes, the volume of water was determined, which can be utilized in the case of the technologies described above, and it will constitute an added value.

1. Introduction

Waste from energy or mining has been widely used in civil engineering for many years, i.e. in road engineering, construction, land reclamation as well as in mining. In mining, waste from energy is of particular technological importance. In Poland, they are often referred to as UPS (Coal Combustion Products) - the equivalent of the global CCP. UPS includes fly ash, ashes with flue gas desulphurization products, flue gas desulphurization products themselves, e.g. reagips, ash-slags, slags and bottom ashes. UPS (if it is a waste and not a product) in accordance with the Regulation of the Minister of Environment of 9 December 2014 on the waste catalog UPS are included in group 10 as waste from thermal processes. Paper [1] describes legal aspects regarding the recovery of energy waste in mining plants. The number of used UPSs in the Polish mining system is systematically decreasing every year [2]. UPS are used in the following underground mining technologies [3]:

- grouting of cavings as fire prevention,
- elimination of unnecessary dog headings,
- filling shallow voids in the rock mass,
- making insulation plugs,
- liquidation of shafts,
- extinguishing fires.

The most frequently performed technology in underground mining with fine-fraction waste is the sealing of infarcted goafs [1]. The technology uses various types of fine-grained waste, in particular fly ash. Fly ash has a diverse chemical, phase and morphological composition, which affects their physico-chemical properties [4]. Fly ash can be divided into two main groups: low-calcium ash (LCFA) and high-calcium ash (HCFA) [5]. Another criterion for the division may be the type of furnace in which they arise. Two main types of furnaces in which hard coal is burnt are: conventional boilers and fluidized bed boilers. Such a division is often used by practitioners from mines. At present, in classic boilers in which hard coal is burnt, ash with the code 10 01 02 arises. They are generally low-calcium ash (LCFA). On the other hand, fluidized bed boilers primarily produce ashes with code 10 01 82, which contain flue gas desulfurization products and are usually highly calcium-refractory (HCFA). These ashes differ mainly in the content of silicates, aluminum, calcium, and in particular its free form and sulfur compounds. Currently, in Polish hard coal mines located in Upper Silesia, the most commonly used technologies are ashes with the code 10 01 02 from conventional boilers, and as second ashes with code 10 02 82 from fluidized bed boilers. As previously mentioned, fly ash was most often used in the technology of sealing the infarcted grouting of cavings and in the elimination of unnecessary dog headings. Transport and migration properties play the first role in these technologies. They depend, inter alia, on the type of fly ash. The article presents the results of research on selected physical properties of ash-water mixtures such as: water-ash index (W/FA), leveling, density and volume of excess water. Furthermore for the tested ashes, the volume of water is determined, which can be utilized in the case of the technologies described above.

2. Material and methods

The tests were carried out for the following wastes:

- fly ash (LCFA) coming from conventional "A" power plant boilers with code 10 01 02,
- fly ash (HCFA) coming from the fluidized bed boiler of the "B" power plant code 10 01 82.

The tested ashes come from selected power plants located in the south of Poland, in which coal is burning.

The research results of physical properties presented in the article were carried out in accordance with the PN-G 11011: 1998 standard. Tests of chemical composition in the oxide form were made using the X-ray fluorescence methode (XRF) using the Epsilon 1 spectrophotometer by PANalytical. In contrast, the chemical composition of the water extract was tested in accordance with PN-EN 12457-4: 2006 standard.

Selected fly ash A and B were mixed with water to form ash-water mixtures. The proportions of fly ash and water (W/FA ratio) were determined in such a way that the resulting mixtures had a liquid consistency, the measure of which is pourability. The liquid consistency results from the fact that the transport of ash-water mixtures from the surface to the underground excavation is carried out by means of pipelines in a gravitational manner. The fluidity of the mixture determines by about their transport and penetration capabilities.

3. Results and discussion

3.1. Chemical properties

The chemical composition in the oxide form of the tested ashes is shown in table 1.

Table 1. Chemical composition in the oxide form of the tested ashes.

Chemical ingredient	Fly ash "A" (from a conventional boiler)	Fly ash "B" (from a fluidized boiler)
	% by mass	% by mass
SiO_2	50.08	37.82
Al_2O_3	25.80	19.63
Fe_2O_3	6.10	5.74
CaO	3.80	18.52
MgO	2.23	1.76

Na_2O	0.97	1.09
K_2O	3.02	2.01
SO_3	0.67	7.91
TiO_2	1.08	0.72
P_2O_5	0.50	0.29
CaO free	0.17	3.88
LOI	4.79	3.84

The tested ash are mainly different in content of SiO_2 and CaO and to a lesser extent Al_2O_3. Ash "A" (conventional fired) contains more silicon and aluminum than ash "B" (fluidized bed boiler). In contrast, ash "B" contains much more calcium, in particular its free form, than ash "A". This is due to the fact that ash "B" comes from a fluidized bed boiler, where dry flue gas desulfurization is used. The content of free lime, sulfur or unburned coal in the ashes increases their demand for water, so their water demand increases [6]. Bearing in mind the guidelines of the American ASTM standard in terms of silicon, aluminum and iron content, fly ash "A" could be classified as class F (LCFA) and fly ash "B" to grade C (HCFA). The tested ashes, however, do not fully meet ASTM requirements in terms of alkali, and in the case of ash B also in terms of sulfur. Such acceptance of classification is quite frequent, as the authors of the paper write about [6].

The results of the leaching tests (1:10) from the tested fly ashes are presented in table 2.

Table 2. The results of the leaching tests (1:10) from the tested fly ashes.

Substance	Fly ash "A" (from a conventional boiler)	Fly ash "B" (from a fluidized boiler)
	mg/l	mg/l
chloride	32	191
sulphate	389	1102
Cr	0.14	0.054
Zn	<0.05	0.050
Cd	<0.001	<0.05
Mg	<0.5	<0.5
Mn	<0.005	<0.005
Cu	<0.005	<0.005
Ni	<0.005	<0.005
Pb	<0.005	<0.005
K	16.85	13.10
Na	29.40	16.85
Fe	<0.005	<0.005

The ash investigated in terms of the chemical composition of water extracts differs mainly in the content of sulphates and chlorides. Fly ash "A" (from a conventional boiler) is characterized by several times lower content of chlorides and sulphates than ash "B" (from a fluidized bed boiler).

3.2. Grain size composition

The grain size composition of the examined ashes was determined by the traditional sieve method. The curves of the grain composition of ash "A" and ash "B" are shown in figure 1.

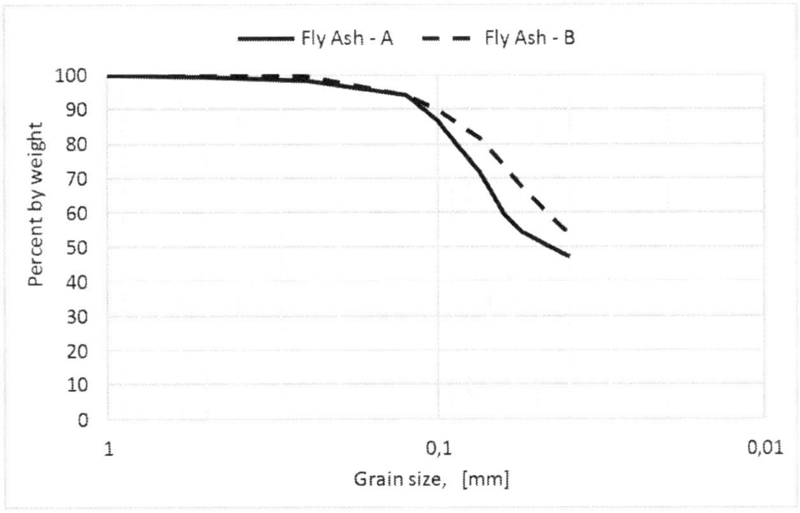

Figure1. Grain composition curves for the examined ashes.

The ash "A" coming from a conventional boiler is characterized by a slightly coarser grain size than the ash "B" coming from a fluidized bed boiler in the tested range up to 0.063 mm, which was also demonstrated, for example, in [7]. However, these ashes significantly differ in the structure of the grain. Ashes from conventional boilers, such as ash "A", are made of spherical grains, which reduces the need for water [8]. In contrast, ashes from fluidized bed boilers ("B" ash) are characterized by fine grains of very irregular shape, which are often compacted together to form agglomerates [9]. This state is reflected in the examined ashes. An exemplary image of grains of fly ash formed from the combustion of hard coal in conventional and fluidal boilers is shown in figure 2 [9].

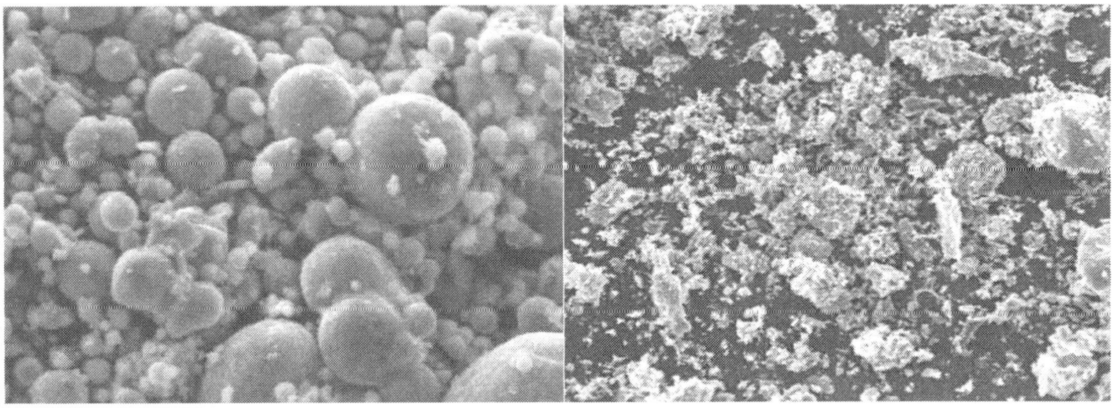

Figure 2. An exemplary image of grains of fly ash resulting from hard coal combustion in conventional and fluidized boilers [9].

The grain size composition of fly ash influences, among others on their water demand [10]. The specific surface of the grain also affects the water demand [11]. The specific surface of fly ash from conventional Polish power plants, i.e. ash "A", according to research carried out by the author of the work [12] is in the range from about 2500 cm^3/g to about 4000 cm^3/g, on average 3083 cm^3/g . However, the specific surface of fly ash from fluidized bed boilers, i.e. ash "B", according to work [12], is very diverse in the range from approximately 4500 cm^3/g to 7000 cm^3/g, on average 5797 cm^3/g. The quoted surface test results indicate that fly ash from fluidized bed boilers is characterized by almost twice as large surface area as ash from conventional boilers. The specific surface also affects the pozzolanic reactivity of fly ash with concrete and cement produced with their participation [13].

3.3. The leveling

The dependence of leveling, which is a measure of the fluidity of mixtures, from the water-and-ash indicator (W/FA) is shown in figure 3.

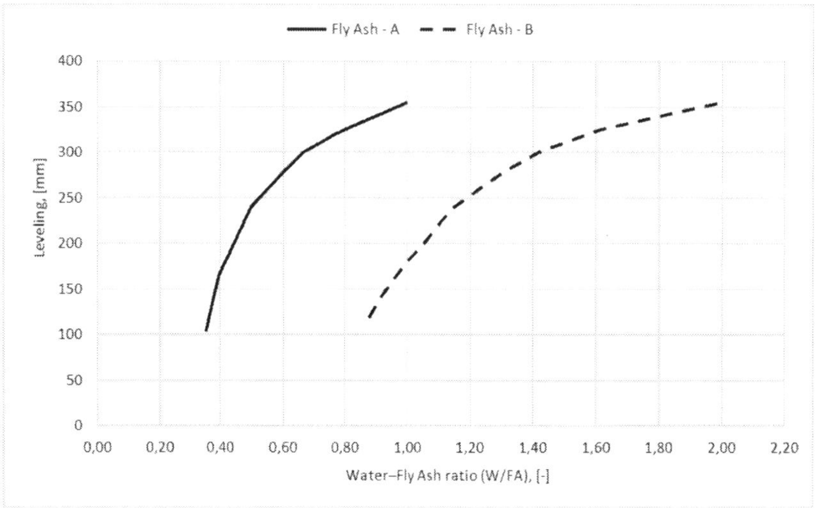

Figure 3. The dependence of the leveling from the water-and-ash indicator (W/FA).

The testing of leveling was carried out in the range of water-ash indicator (W/FA) from 0.35 to 1.00 for ash "A" and in the range from 0.88 to 2.00 for ash "B" (FCB). This range of water and ash indicators made it possible to reach leveling in a very wide range from 100 mm to 355 mm. The conducted research clearly shows the strong dissimilarity of the examined ashes in terms of water demand for a given level of leveling. Fly ash "B" coming from a fluidized bed boiler is much more water than the fly ash "A" coming from a conventional boiler. The reason for this is primarily the different grain shape of the fly tested ash, as well as the differences in the chemical composition, in particular in the content of SiO2 and CaO, which has been described above. For example, for 160 mm water level, the water-ash indicator for ash "B" is 2.4 times higher than for ash "A". However, for the 355 mm leveling, the quotient decreases to 2.00. In the range of 160 mm to 355 mm leveling, the average value of the ratios "B" to ash "A" is 2.25. This is reflected in the specific surface, which for ash from fluidized bed boilers (ash "B") is about twice as large as ash from conventional boilers (ash "A"). A further reason for the difference in the water demand of the tested ashes is related to their chemical composition, which has been described above. The water-ash indicator ratio (from a fluidized bed boiler) to ash "A" (from a conventional boiler) for selected fillings are presented in figure 4.

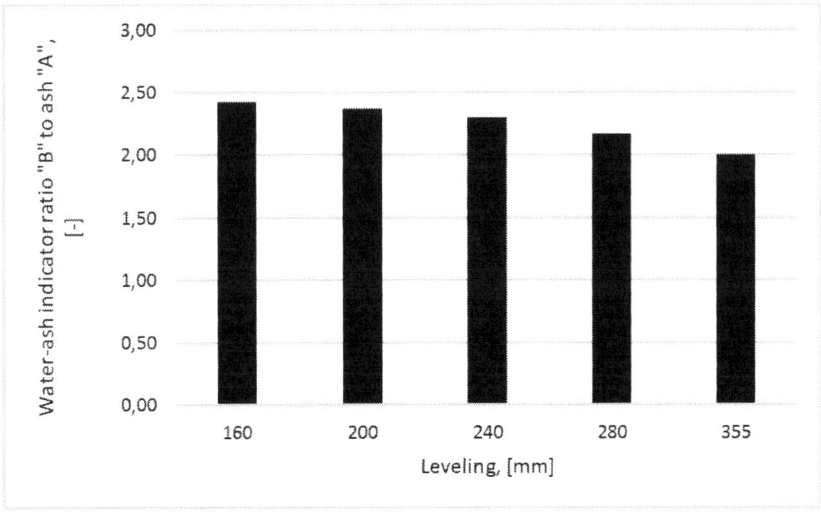

Figure 4. Water-ash indicator ratio ash "B" to ash "A" for selected leveling.

3.4. Density of mixtures

The dependence of the density of the water mixtures of the ashes from the leveling is presented in figure 5.

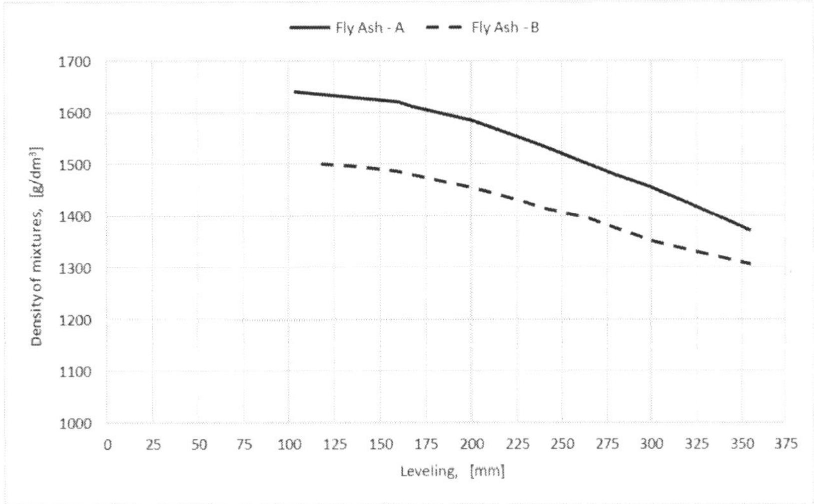

Figure 5. Dependence of the density of water mixtures of the ashes from the leveling.

The fixed proportions of water to ash (W/FA index) and the resulting leveling rate determine the density of the obtained mixtures. The density of mixtures produced with the ash "A" in the range from 103 mm to 355 mm is from 1370 g/dm^3 to 1640 g/dm^3, respectively. However, the density of mixtures with the share of ash "B" is from 1295 g/dm^3 to 1500 g/dm^3 corresponding to the leveling range from 118 mm to 355 mm. The conducted studies show the diversity of the density of water mixtures of the tested ashes. This is mainly due to the fact that the ashes tested largely differ in the water demand, which is expressed by the W/FA index, as previously described, and the reason for which is the diversified grain shape and chemical composition, as previously discussed. The density of mixtures made of ash "A" is higher than for mixtures with ash "B". Depending on the leveling, this is from

IOP Conf. Series: Materials Science and Engineering **545** (2019) 012004 doi:10.1088/1757-899X/545/1/012004

about 9% for leveling in the limit of 110 mm, up to around 5%, for leveling of about 350 mm. The quotient of water mixtures density ash "A" (from a conventional boiler) to the ash "B" (from a fluidized bed boiler) for selected levels is shown in figure 6. In the range of 160 mm to 355 mm leveling, the average value of the ash density quotients "B" up to ash „A" is 1.078, or about 8%.

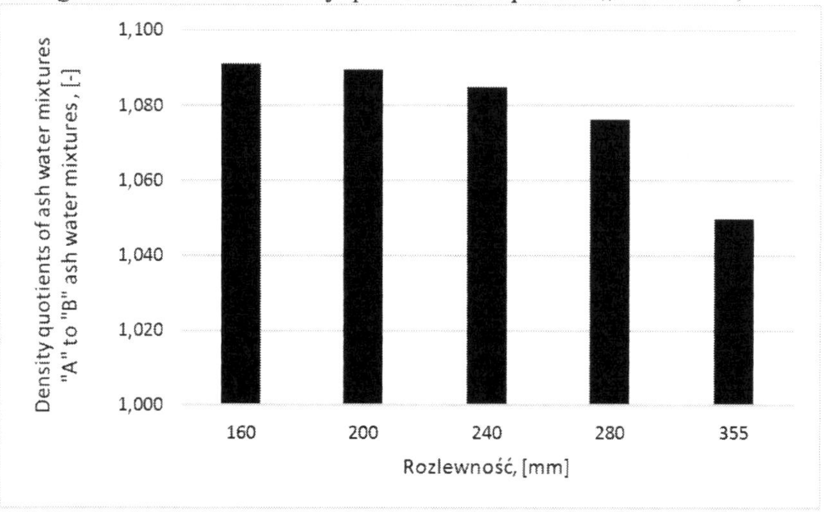

Figure 6. Density quotients of ash water mixtures "A" to "B" ash water mixtures for selected leveling.

3.5. The volume of excess water

Excess water is the water that accumulates over the sediment. It is undesirable because it has to be discharged. Usually it goes to the mine drainage system, generating unnecessary costs associated with its pumping to the surface. However, mines very often use mixtures with significant leveling at 300-350 mm. This is dictated by significant transport distances and the lack of possibility to obtain a full power supply, i.e. with a hydraulic height equal to the distance from the surface to the place of application. The basic factor of the inability to get a full power supply is the diameter of the used pipelines. In some cases, it may also be related to the inability to produce a mixture with a low water and ashes index (W/FA) or to obtain adequate capacity in the generating installation. The dependence of the volume of excess water from the tested ashes on the leveling is presented in figure 7.

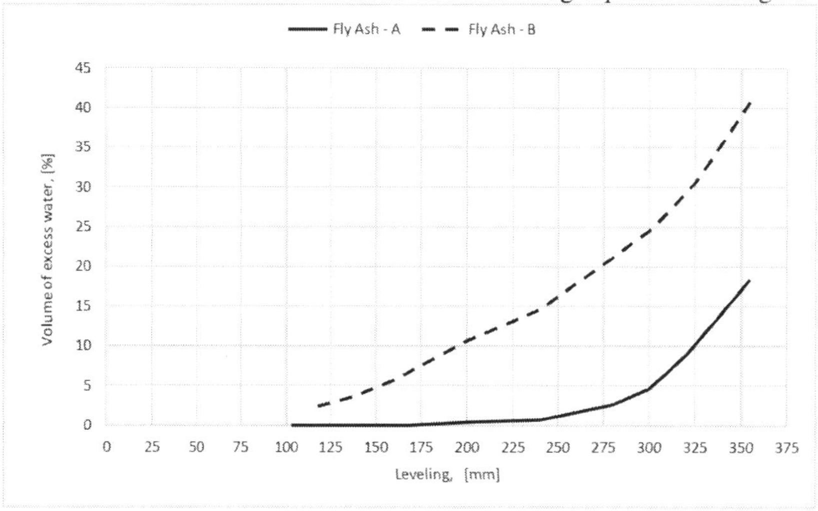

Figure 7. Dependence of the volume of excess water from the tested ashes.

The water volume of excess mixtures produced with ash "A" in the leveling range from 103 mm to 355 mm is from 0% to about 18%, respectively. It should be noted that ash water mixtures "A" practically do not exceed about 170 mm, they do not give off excess water. It is reflected in the work [14]. However, this can not be said for mixtures made of ash "B" (from a fluidized bed boiler). Analyzing further the course of the variability of excess water exudation from the "A" ash mixtures, it should be noted that for the leveling of about 250 mm its volume is negligible (less than 1%). A significant increase in excess water release can only be observed from the 300 mm level, at which the mixture gives off about 5%. The increase in leveling from 300 mm to 350 mm, i.e. by 50 mm, causes an increase in the volume of excess water to about 18% - a fourfold increase.

The mixture made from ash "B" (from a fluidized bed boiler) gives off much more excess water than the above-described mixtures made of ash "A" (from a conventional boiler). The water volume of the excess mixtures produced with ash "B" in the flow range from 118 mm to 355 mm is suitably from about 2.5% to about 41%. Ash water mixtures "B" are characterized by lack of leveling for which they would not give off excess water. Excess water for ash mixtures "B" increases with the increase in the W/FA ratio. The quantities of the excess water volume of the ash "B" mixtures (from the fluidized bed boiler) to the ash mixtures "A" (from the conventional boiler) for the selected leveling are shown in figure 8.

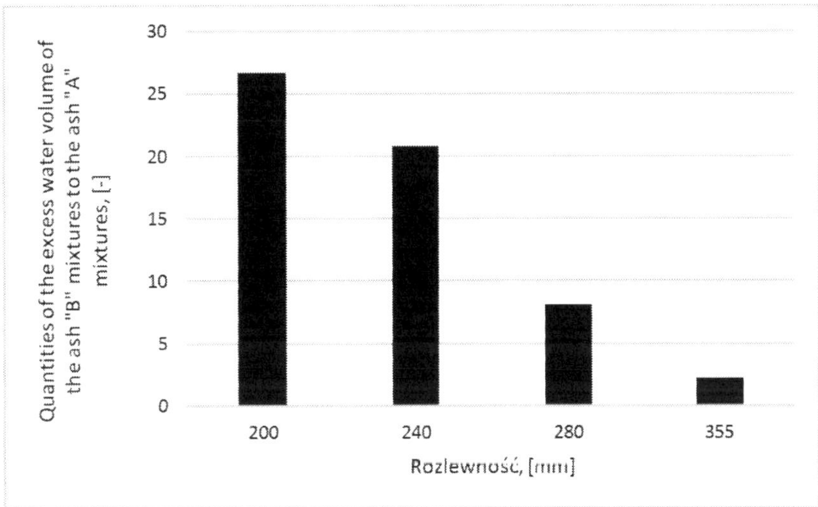

Figure 8. Quantities of the excess water volume of the ash "B" mixtures to the ash "A" mixtures for selected leveling.

The quotient presented in figure 8 illustrates the very strong differentiation of the volume of excess water volume for the tested ashes depending on the level of flow. The greater the flow rate, the smaller the quotient. For example, for 200 mm pourability, the volume of excess water released by the mixture of ash "B" (from the fluidized bed boiler) is about 27 times higher than for the ash mixture "A" (from a conventional boiler). However, for the 355 mm leveling, the quotient decreases to about 2. In the flow range from 160 mm to 355 mm, the average value of the quotients of the amount of excess ash "B" to ash "A" is about 14.

3.6. The volume of water that can be absorbed

Taking into account the water and ash indicator (W/FA) and the results of the density and volume tests of excess water, you can calculate the volume of water that can be absorbed by ash. The volume of water that can be bound by the mass of 1 Mg of the tested ashes depending on the leveling is shown in figure 9.

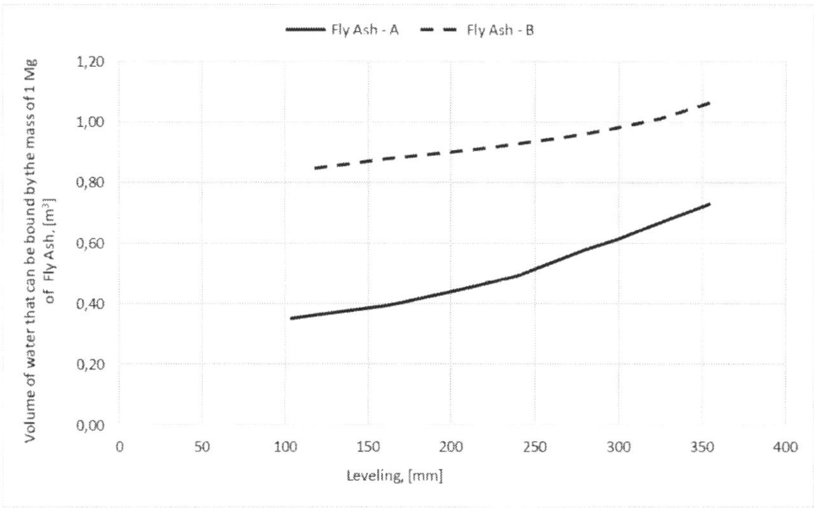

Figure 9. The volume of water that can be bound by the mass of 1 Mg of the tested ashes depending on the leveling.

The ash "A" (from a conventional boiler) allows, in the leveling range from about 100 mm to about 350 mm, the water absorption from about 0.35 m^3 to about 0.75 m^3 with respect to 1 Mg of dry matter. On the other hand, 1 Mg of ash "B" (from a fluidized bed boiler) allows absorption of about 0.8 m^3 to about 1.1 m^3 of water in the range of tested leveling. The quotient of the water volume that can be absorbed by the water mixtures of ash "B" into ash "A" is shown in figure 10.

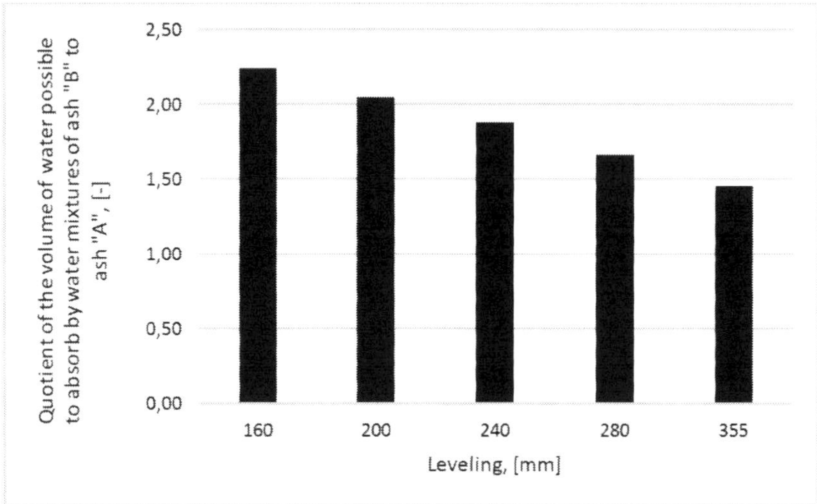

Figure 10. The quotient of the volume of water possible to absorb by water mixtures of ash "B" to ash "A".

At 160 mm, the ash "B" absorbs about 2.2 times more water than the ash "A", and at a flow rate of 355 mm about 1.5 times more. The average value of the quotient possible to absorb the volume of ash water "B" to ash "A" in the range of tested leveling is about 1.9. In general, ash "B" coming from a fluidized bed boiler allows the management of almost twice as much volume of mine water.

4. Conclusions

- Fly ash "B" (from a fluidized bed boiler) has a 2.4-2.0 times higher water-ash ratio (W/FA) compared to ash "A" (from a conventional boiler) with a leveling range of 160 mm, respectively up to 355 mm, average 2.25.
- Despite the significant differences in the W/FA ratio, the density of the water mixtures of the tested ashes is similar, and their difference varies from about 9% to about 5% for leveling rates from about 110 mm to 355 mm, respectively.
- Ash water mixtures "A" emit much less excess water than "B" ash. The volume of superfluous excess water is heavily dependent on the W/FA ratio, and hence on the level of leveling. With the leveling often used by mines of 280-350 mm, the volume of excess water released ranges from about 2.5% to about 18% for ash "A and from about 20% to about 40% for as "B".
- Despite the large volume of excess water released by ash "B" in comparison to ash "A", it must be stated that ash from fluidized bed boilers allows the management of almost twice the volume of unnecessary mine water, compared to ash from conventional boilers.

References

[1] Pierzyna P 2017 *IOP Conf. Ser. Mater. Sci. Eng.* **268** p 012011
[2] Drobek L, Borecki T and Pierzyna P 2015 *Wiad. Gór.* **66**(5)
[3] Drobek L, Borecki T and Pierzyna P 2015 *Wiad. Gór.* **66**(6)
[4] Szponder D K and Trybalski K 2009 *Gór. Geoinżynieria* **33**(4) pp 287–98
[5] Giergiczny Z 2002 Popiół lotny aktywnym składnikiem cementu 4 *Sympozjum Naukowo-Techniczne: Reologia w technologii betonu* pp 5–17
[6] Tsimas S and Moutsatsou-Tsima A 2005 *Cem. Concr. Compos* **27** pp 231–7
[7] Tomaszewicz M and Zuwała J 2015 *Przem. Chem.* **94(4)** pp 488–95
[8] Felekoglu B 2006 *Fuel* **85** pp 1944–9
[9] Rajczyk K 2012 Popioły lotne z kotłów fluidalnych i możliwości ich uszlachetniania, *Opole: Wydawnictwo Instytut Śląski*
[10] Zapotoczna-Sytek G 2013 Popioły lotne nowej generacji do produkcji autoklawizowanego betonu komórkowego: monografia *Warszawa; Opole: Wydawnictwo Instytut Śląski*
[11] Felekoğlu B, Türkel S and Kalyoncu H 2009 *Constr. Build. Mater.* **23** pp 2053–61
[12] Szponder-Kołakowska D K 2012 Badania wybranych właściwości popiołów lotnych z zastosowaniem analizy obrazu, *rozprawa doktorska*
[13] Wang T, Ishida T and Gu R 2018 *Constr. Build. Mater.* **190** 1163–72
[14] Piotrowski Z and Mazurkiewicz M 2006 *Gór.Geoinżynieria* **30**(3) 37–46

Measurements of dust concentration at workplaces in a hard coal mine processing plants after installation of the NEPTUN spraying system

D Bałaga[1], M Kalita[1], M Siegmund[1] and Z Klimek[2]

[1]KOMAG Institute of Mining Technology, Division of Machines and Equipment, 37 Pszczyńska Street, 44-101 Gliwice, Poland
[2]KWK Bolesław Śmiały in Łaziska, 12 Świętej Barbary Street, 43-170 Łaziska Górne, Poland

dbalaga@komag.eu

Abstract. The article presents the place and scope of testing the NEPTUN prototype spraying system intended for neutralization of total and respirable dust at workplaces in the processing plant of Bolesław Śmiały mine. The testing procedure and the results of dust concentration measurements at each workplace, covered by the operational range of the NEPTUN spraying system are presented. In the final part, the results of tests were analyzed and the reduction efficiency of total and respirable dust at the tested workplaces is given.

1. Introduction

Exploitation of hard coal deposits exposes workers to technical and natural hazards. In addition to natural hazards such as methane, rock bursts, floods, fires, gas and rock outbursts as well as climate hazard, a treat associated with the presence of coal dust, should be mentioned. Due to hard coal mine activities aimed at increasing the work safety as well as employees awareness, no coal dust explosion has been reported for over a decade [7]. However, the presence of coal dust in the mine atmosphere above the maximum permissible concentrations (NDS), exposes the respiratory system of workers to dust, causing pneumoconiosis, i.e. an incurable occupational disease. Despite the use of the state-of-the-art methods and measures to control the dust [1] [2], there are still several hundred cases of new cases of pneumoconiosis per year (Figure 1) [7].

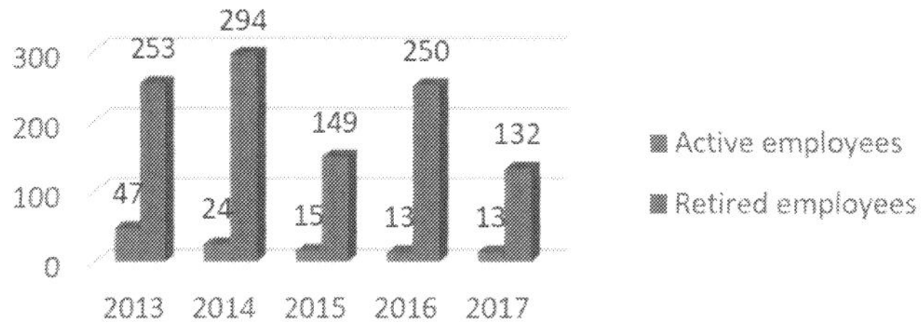

Figure 1. Number of pneumoconiosis cases observed within the years 2013-2017 for employees of operating coal mines [7].

Main sources of mine airborne dust are related to mining the coal wall by shearers, development of roadway workings, moving the powered roof supports units and crushing and transporting the run-of-mine (especially at the chutes) [4]. Hard coal processing plants are the places in hard coal mines exposed to high concentrations of dust, where the following processes: classification, grinding and beneficiation take place. The main sources of dust include: preliminary screens, along with transfer and receiving shafts, pre-cut crushers, pre-classification screens with vibratory feeders or transfer shafts, non-metallic contaminant removal tapes and iron trapping, or crusher for selective separation of coarse-grained waste [6]. The paper [5] presents the results of analyses of coal dust concentration at workplaces in the processing plants of all the Polish mines. They showed that almost in all the processing plants there were the workplaces with dust concentration over the MAC (Figure 2).

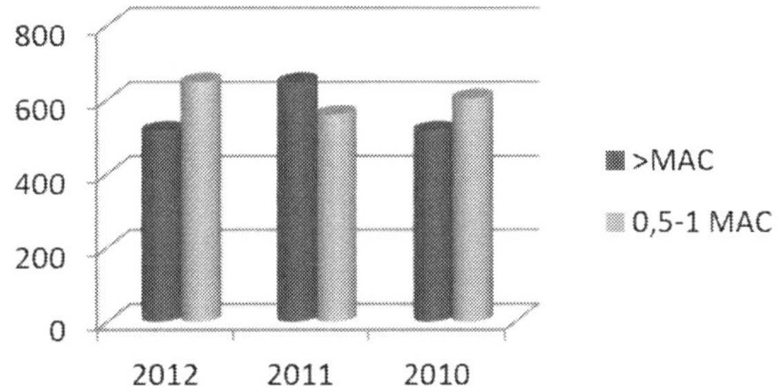

Figure 2. Total number of workplaces in hard coal mines processing plants threatened by dust hazard between
2010 and 2012 [5].

The presented results clearly showed that further work was needed to reduce the amount of dust generated at workplaces in the processing plants. The NEPTUN spraying system, applied in the Mechanical Processing Plant of Polska Grupa Górnicza sp. z.o.o. Bolesław Śmiały Mine is the latest solution, designed at KOMAG, for a reduction of dust concentration at workplaces below MAC [3]. The NEPTUN spraying system is designed for a precipitation and elimination of dust particles emitted in the result of feeding the material in the area of belt conveyors, chutes and screens, using water drops produced by spraying nozzles (Figure 3) [5].

Figure 3. Spraying stream forming mist for a precipitation of dust particles from air on one of the protected workplaces [9].

A task of the system, installed in the processing plant, is to reduce dust concentration at workplaces to the level required by the standards for inhalable and respirable dust, in accordance with the

IOP Conf. Series: Materials Science and Engineering **545** (2019) 012005 doi:10.1088/1757-899X/545/1/012005

regulation of the Minister of Family, Labor and Social Policy of 12 June 2018 [8]. PN-Z 04008-7 of August 2002 Standard was used to develop the testing methodology, on air purity protection and principles of air sampling in the work environment as well as interpretation of results. The research work was aimed at verifying the correctness and effeciency of the NEPTUN system in reduction of dust hazard.

2. Place and scope of tests

The NEPTUN spraying system solution for a dust reduction in a processing plant consists of seven independent spraying installations in five places [3]:

- tanks, loading and unloading devices for coal and stone,
- beneficiation devices,
- crushing and transporting devices,
- tanks, coal loading and unloading equipment,
- feeders at the level of +3.70 m.

The NEPTUN spraying system uses water drops with the appropriate size and energy characteristics for a precipitation of dust particles. The spraying system uses special ultrasonic atomizers, in which compressed air is an additional medium. They enable to operate in the low pressure range of water and compressed air, creating a spraying stream, which does not significantly increase water content in the transported material. For such protected workplaces, dust concentration measurements were taken to determine the dust reduction efficiency. The tests were carried out for properly working spraying installations and to compare the results, with the switched off installation. The people, working on the stations with the NEPTUN installation, were equipped with measuring devices. Dust concentration measurements were taken with the following two types of gravimetric personal dust meters:

- CIP-10R for measuring the respirable dust fraction,
- CIP-10I for measuring the inhalable dust fraction.

Amounts of dust from the measuring cups of personal dust meters were weighted by the employees of the Laboratory of Air Pollution Measurement of the Central Mining Institute in Katowice, which is a body accredited by the Polish Center for Accreditation (PCA) in Warsaw in the scope of environmental tests.

Amounts of dust collected at each workplace enabled to determine the inhalable and respirable dust concentration in the conditions of operating spraying installations, and then to compare them with the results of inhalable and respirable dust concentration when the spraying installation was switched off. This allowed to determine the effectiveness of the installed NEPTUN spraying system.

3. Testing procedure

After starting the NEPTUN spraying system, employees working at each workplace were equipped with the measuring devices. The person equipped with the measuring device, consisting of a dust meter to determine the inhalable and respirable dust concentration went to the workplace with operating spraying system (Figure 4).

Figure 4. NEPTUN installation for a reduction of dust concentration at one of the protected dust generating points [9].

The people equipped with the measuring devices were instructed about the need to perform activities being within the scope of their duties and checking the signaling lamps on the devices confirming their correct operation. The personal dust meters (Fig. 5) CIP-10I (for measuring the inhalable dust fraction) and CIP-10R (for the measurement of the respirable fraction) were placed on the chest. These devices took air samples for a set time with an air output equal to a human demand during a typical activity (about 10 l / min).

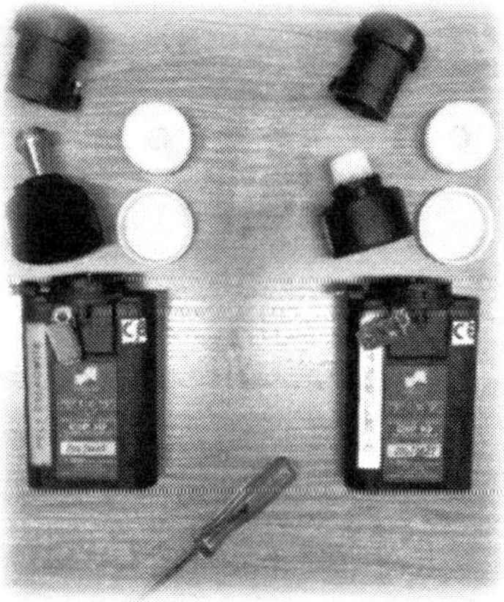

Figure 5. Personal dust meter (on the left CIP-10I for measuring the inhalable fraction, and on the right CIP-10R for measuring the respirable fraction [9]

Twenty measuring cups were used for a collection of respirable and inhalable dust. The cups were prepared and marked before the measurement, so it was possible to identify the workplace from which the dust samples were taken. The tests were conducted for four hours. Similarly, dust concentration measurements were taken with the NEPTUN spraying system switched off, with the air sampling time of about 1 hour. The shortened test time with the spraying system switched off was dictated by the need to limit the exposure of employees to harmful conditions. After the tests, the measuring cups were removed from the dust meters and protected against unwanted dust spillage out. The cups

IOP Conf. Series: Materials Science and Engineering **545** (2019) 012005 doi:10.1088/1757-899X/545/1/012005

secured in this way were transported to the laboratory in which they were subjected to the following procedures:

- drying at 50-60° C for two hours,
- seasoning in the desiccator over a period of 3 hours,
- weighing.

Dust concentration measurements at the Mechanical Processing Plant of Polska Grupa Górnicza S.A., Bolesław Śmiały Coal Mine, enabled to determine the inhalable and respirable dust concentrations at each workplace with the dust control installation switched on and off.

4. Results of measurements

Dust amounts, collected in the cups during the tests, are presented in Table 1, which contains dust amounts obtained from the measurement of the inhalable and respirable fraction in each workplace, with the operating NEPTUN dust reduction system and the measurement time. In turn, Table 2 presents the dust amounts obtained from the measurements of the inhalable and respirable dust fraction in individual work stations, with the NEPTUN installation switched off together with the measurement time.

Table 1. Dust amounts, measurement time and measuring cup marking for each workplace, when the NEPTUN spraying system is switched on.

Workplace	Type of dust fraction	No. of cup	Dust weight in the cup [mg]	Measurement time [min]
I	respirable	8k	2.59	215
	inhalable	10k	12.73	
II	respirable	2k	1.1	210
	inhalable	1k	4.95	
IIIa	respirable	9k	0.45	210
	inhalable	6k	2.43	
IIIb	respirable	3k	0.24	195
	inhalable	7k	2.86	
IV	respirable	5k	1.3	195
	inhalable	4k	3.72	

Table 2. Dust amounts, measurement time and measuring cup marking for each workplace, when the NEPTUN system is switched off.

Workplace	Type of dust fraction	No. of cup	Dust weight in the cup [mg]	Measurement time [min]
I	respirable	157k	1.21	60
	inhalable	178k	7.13	
II	respirable	160k	1.32	64
	inhalable	158k	3.47	
IIIa	respirable	154k	0.34	60
	inhalable	167k	1.87	
IIIb	respirable	171k	0.59	65
	inhalable	183k	3.55	
IV	respirable	187k	0.84	62
	inhalable	156k	2.47	

The dust amounts from dust concentration measurements with the NEPTUN spraying system switched on and off enabled to determine the average respirable and inhalable dust concentration at the selected workplaces. To determine the concentration of dust in the air, the following relationship was used:

$$\bar{S} = \frac{m}{\dot{V} * \tau} \tag{1}$$

where:

$\bar{S}$ – dust concentration at the measurement site [mg/m^3]
$\dot{V}$ – air flow rate in CIP 10 [dm^3/min]
τ – measurement time [min]
m – weight of dust in the measuring cup after the exposure time τ [mg]

The calculation results of inhalable and respirable dust concentrations at the tested workplaces, for the NEPTUN spraying system switched on and off, are presented in Table 3.

Table 3. The average inhalable and respirable dust concentration obtained during tests at each workplace with the NEPTUN spraying system switched on and off.

Workplace	Type of dust fraction	Average dust concentration - installation switched on [mg/m^3]	Average dust concentration - installation switched off [mg/m^3]
I	respirable	1.20	2.02
	inhalable	5.92	11.88
II	respirable	0.52	2.06
	inhalable	2.36	5.42
IIIa	respirable	0.21	0.57
	inhalable	1.16	3.12
IIIb	respirable	0.12	0.91
	inhalable	1.47	5.46
IV	respirable	0.67	1.35
	inhalable	1.91	3.98

For these inhalable and respirable dust concentrations, with the NEPTUN spraying system switched on and off, at each workplace, it was possible to determine the efficiency of dust reduction. The dust reduction efficiency is defined as the quotient of the difference between the concentration of dust (inhaled and respirable) measured with the spraying system switched off and the concentration of dust (inhaled or respirable) measured with the spraying system switched on and the concentration measured with the spraying system switched off.

The percentage efficiency of dust reduction was determined from the following formula:

$$\eta_{red\ zap} = \frac{\bar{S}_1 - \bar{S}_2}{\bar{S}_1} x100\% \tag{2}$$

where:

$\bar{S}_1$ – dust concentration with the spraying system switched off
$\bar{S}_2$ – dust concentration with the spraying system switched on

The determined dust reduction efficiency, resulting from the use of NEPTUN dust reduction equipment at each workplace, is shown in Table 4.

Table 4. Dust reduction efficiency (inhalable and respirable dust fraction) obtained at individual workplaces with use of the NEPTUN spraying system.

Workplace	Type of dust fraction	Efficiency of dust reduction [%]
I	respirable	40
	inhalable	50
II	respirable	75
	inhalable	57
IIIa	respirable	62
	inhalable	63
IIIb	respirable	86
	inhalable	73
IV	respirable	51
	inhalable	52

5. Analysis of results

On the basis of the tests carried out and of the results of inhalable and respirable dust concentration at each workplace, after the NEPTUN spraying system was installed, there were no exceeded values in maximum allowable dust concentrations, which for inhalable dust reach 10 mg/m^3, and for respirable dust up to 2 mg/m^3 [8]. (Figure 6)

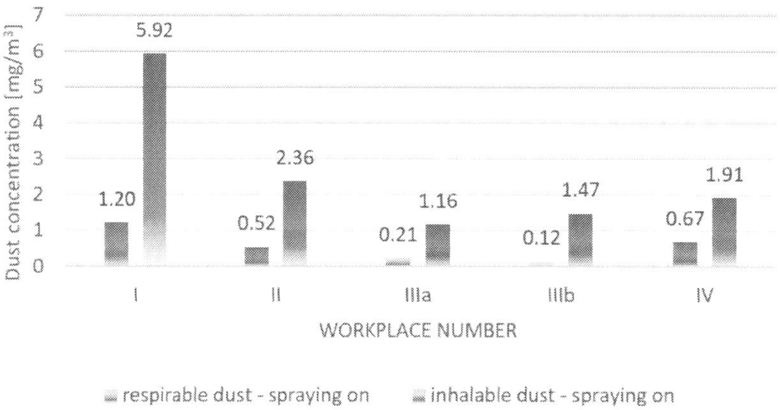

Figure 6. Concentrations of inhalable and respirable dust measured at workplaces with the NEPTUN Spraying System switched on [9].

At the workplaces No. II, IIIa, IIIb and IV protected by the spraying installation, low concentration of inhalable and respirable dust can be observed, not exceeding 3 mg/m^3 for inhalable dust which is 30% of MAC and 1 mg/m^3 for respirable dust which is 50% of MAC. In the case of workplace No. I, there is a high disproportion in the dust concentration results, which in the worst case is five times the lowest measured concentration at the workplace No. IIIa. Stone dust from crushers located near the workplace No. I has an impact on inhalable dust (~6 mg/m^3) and respirable dust (~1.2 mg/m^3) concentration (60% of MAC for inhalable and respirable dust). The places of installation of crushers were not covered by the spraying system. Dust concentration measurements with the dust reduction installation switched off, despite the short measurement time, showed the exceeded value of the Maximum Allowable Concentrations. Exceeding of MAC for inhalable and respirable dust was observed at workplace No. I and No. II. The inhalable and respirable dust concentrations at each workplace with the NEPTUN system switched off are shown in Figure 7.

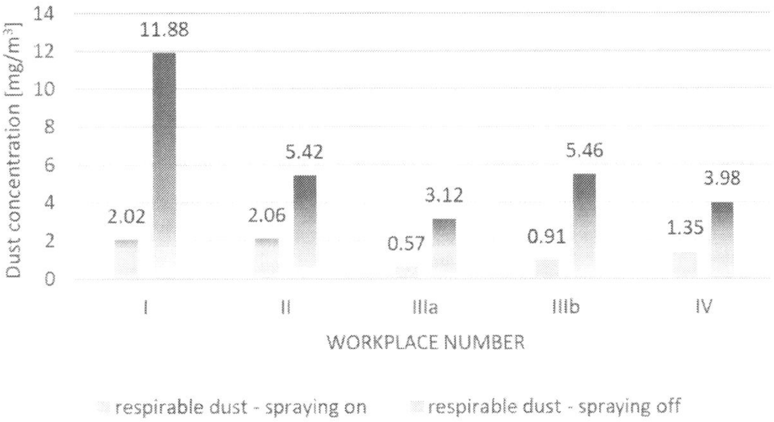

Figure 7. Inhalable and respirable dust concentration measured at the workplaces with the NEPTUN Spraying System switched off [9].

The lowest dust reduction efficiency 50% was obtained at the workplace No. I, and the highest one at the workplace IIIa (63%) and IIIb (73%) (Figure 8).

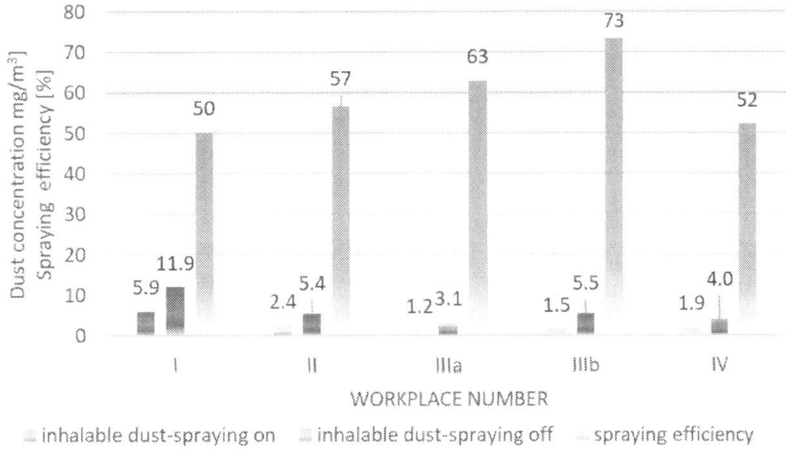

Figure 8. Efficiency of dust reduction at individual workplaces, with the NEPTUN Spraying System [9].

When comparing the results of the respirable dust concentration measured with the operating spraying system and the results of the respirable dust concentration measured with the spraying system switched off, the worst dust reduction efficiency, about 40%, was detected at the workplace I (Figure 9).

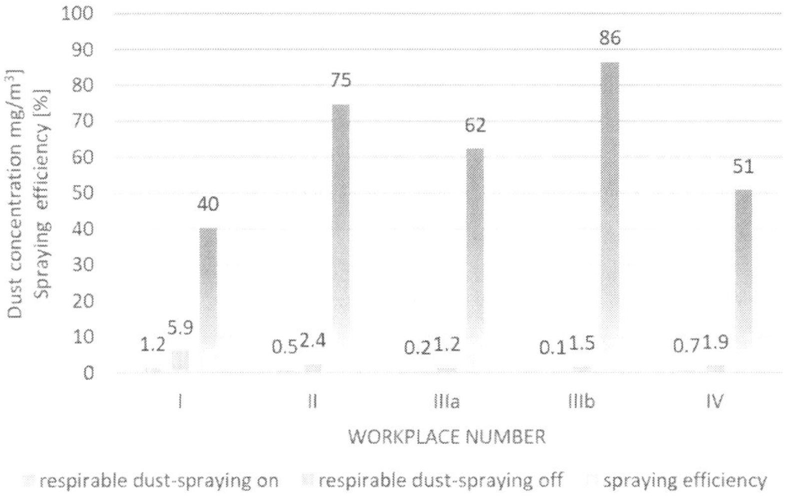

Figure 9. Efficiency of respirable dust reduction at individual work places, with the NEPTUN Spraying System [9]

6. Conclusions

The NEPTUN spraying system presented in the article, developed at the KOMAG Institute of Mining Technology, is designed to reduce a concentration of dust particles by precipitating them with use of water drops generated by air-water atomizers. The system was manufactured and installed at the Mechanical Coal Processing Plant of Polska Grupa Górnicza S.A. "Bolesław Śmiały" Coal Mine by the Elektron Innovation and Implementation Company. The measurements of inhalable and respirable dust concentrations at each workplace clearly showed a positive effect in reduction of dust concentration below the MAC by the NEPTUN Spraying System. In assessment of efficiency, despite unfavourable conditions (lower efficiency of the feed during the measurement of dust concentration with the spraying system turned off), showed at least 50% effectiveness in the case of inhalable dust reduction (workplace I), reaching even more than 70% efficiency at the workplace IIIb. In turn, the efficiency of the spraying system in the reduction of respirable dust was at least 40% (site I - where the stone crusher was unprotected by the system), reaching even 86% efficiency in the case of IIIb. In other cases, the efficiency of respirable dust reduction (apart from position IV, where efficiency was 51%) reached 60%. The highest respirable dust reduction efficiency by the spraying system was achieved at the workplace II (75%) and IIIb (86%). The presented results indicate that with a properly designed spraying system for a precipitation and catching dust particles by water drops of proper size, it is possible to reduce dust concentration below MAC, with a small water consumption below 10 dm^3/min/atomizer. Use of air-water spraying system allows for a significant reduction of dust concentration at workplaces in Mechanical Coal Processing Plants. It is assumed that a similar efficiency would be achieved by a similar system installed in coal feeding systems of coking plants or combined thermal and power plants.

References

[1] Bałaga D, Cebula D, Kalita M and Siegmund M 2015 *Wiad. Gór.* **11** p. 560-568
[2] Bałaga D, Jedziniak M, Kalita M, Siegmund M and Szkudlarek Z 2015 *Masz. Gór.* **3** p. 68-81
[3] Bałaga D, Kalita M, Siegmund M, Klimek Z, Urbanek A and Waloszczyk A 2018 *Masz. Gór.* **3** p. 35-46
[4] Kalus A 2009 *Zagrożenie pyłami szkodliwymi dla zdrowia w górnictwie węglowym* WUG brochure, Katowice

IOP Conf. Series: Materials Science and Engineering **545** (2019) 012005 doi:10.1088/1757-899X/545/1/012005

[5] Lutyński A 2016 *Zagrożenia metanem, pyłem i hałasem występujące w zakładach przeróbki mechanicznej kopalń węgla kamiennego* (Gliwice: Instytut Techniki Górniczej KOMAG) ISBN 978-83-60708-96-5 Series: Innowacyjne techniki i technologie mechanizacyjne p 106

[6] Śmiejek Z 1999 Punktowe i przestrzenne odpylanie węzłów suchej przeróbki węgla kamiennego *Proceedings: Zwalczanie zagrożeń pyłowych w górnictwie* Wisła 5-7 October 1999

[7] Wyższy Urząd Górniczy *Ocena stanu bezpieczeństwa pracy, ratownictwa górniczego oraz bezpieczeństwa powszechnego w związku z działalnością górniczo-geologiczną w 2017 roku (porównanie od roku 2013)* Katowice 2018

[8] Regulation of the Minister of Family, Labour and Social Policy of June 12, 2018 regarding the highest permissible concentrations and intensities of factors harmful to health in the work environment Journal of Laws, item 1286.

[9] Archive of photographic documentation of ITG KOMAG 2018.

IOP Conf. Series: Materials Science and Engineering **545** (2019) 012006 doi:10.1088/1757-899X/545/1/012006

Use of a Jig Beneficiation Process for Obtaining Mineral Raw Materials

D Kowol and P Matusiak

KOMAG Institute of Mining Technology, Division of Preparation Systems,
37 Pszczyńska Street, 44-101 Gliwice, Poland

dkowol@komag.eu, pmatusiak@komag.eu

Abstract. The article presents a construction and operational principle of a pulsating jig, a device for obtaining minerals in a pulsating water medium. A smart control system of the device is discussed. Examples of jig implementations for obtaining minerals from natural deposits and for a recovery of minerals from mine heaps are presented. Tests results of used equipment are provided, indicating a possibility of obtaining purified aggregate and steam coal concentrate of high quality parameters.

1. Introduction

Treatment of hard coal is an integral and extremely important part of the production process of commercial products. Its aim is to increase the quality of extracted raw material and to obtain quality and quantity parameters of final products required by recipients. The quality of coal is improved by subjecting it to a series of technological operations, including a beneficiation in specialized devices [1,2]. Among the many methods used for the material classification, the gravitational separation technologies carried out in air classifiers (e.g. shaking tables, jigs) [4,5], water classifiers (e.g. jigs, hydro-cyclones) [6] and classifiers with heavy medium suspension (e.g. HMC, HMB) [7,8] are especially important.

The KOMAG Institute of Mining Technology carries out research and development work, which results in the technical documentation of machinery and equipment as well as in the results of tests conducted in industrial conditions and in mining and laboratory facilities at its own test rigs.

An example of this activity includes KOMAG pulsating jigs for material beneficiation according to the density difference, used in many coal and other mineral raw materials processing plants. The solutions, used in them, are subject to continuous improvement processes in order to increase a separation efficiency and maintain the reliability of equipment operation [3].

Basing on a many-year experience of the KOMAG Institute of Mining Technology in the construction and technological adaptation of pulsating jigs for a beneficiation of hard coal, a jig to obtain gravel and sand with a simultaneous release of organic and mineral pollutants has been developed.

The result of this work was a pulsating classifier, an innovative solution allowing for a production of sand and gravel of the required quality and an adjustment of technological parameters of products to the market requirements [9,10].

The further modifications enabled designing the device extending the scope of using the classifiers for recovering mineral resources from mine heaps [11-15].

Presenting the possibility of obtaining high-quality raw minerals from the plants processing the aggregates and mine heaps with the use of a pulsating classifier - a state-of-the-art device designed at the KOMAG Institute of Mining Technology, is the paper objective.

2. Construction and operational principle of pulsating classifier

A principle of a classifier operation is based on a typical process of gravitational beneficiation of minerals by stratification in pulsating water of correctly prepared feed, according to its grain composition and density of ingredients [9,16].

The basic component of the pulsating classifier is a set of working chambers. In its upper part, a sieve plate is mounted on which a movement and separation of the material under beneficiation takes place.

Another component of the device is a rotary receiver equipped with a higher density product scraper. Above the receiver there is an overflow chute for a lower density product.

The pulsating movement is caused by a pneumatic system consisting of a blower, a working air tank and pulsatory plate valves, as well as a compressor supplying pulsatory valves and a working water supply system.

The structure is supplemented by feed and product of a higher density chutes as well as a water drainage and material passing through the sieve openings, from the outflow openings, situated in the lower part of the classifier's working chambers.

A classifier is usually installed on a mobile base figure 1, with a service platform and stairs, or it is installed on a structure together with cooperating devices [17].

Figure 1. K-101 pulsating jig on a mobile base [17].

A correct operation of a classifier is ensured by an electronic control system that controls water pulsation depending on technological parameters of the feed and controls a rejection of products by means of electrical pulses provided by a float sensor [17].

KOMAG has designed a series of pulsating classifiers enabling to adapt them to the user's required capacity. The basic parameters of KOMAG classifiers are shown in table 1.

Table 1. Main technical parameters of KOMAG classifiers [13,18].

Classifier type		K-60	K-80	K-100 K-101	K-102	K-150	K-200
Nominal capacity	(t/h)	60	80	100	100	150	200
Total work surface	(m^2)	ca. 2.0	ca. 4.0	ca. 4.0	ca. 4.0	ca. 4.0	ca. 4.0
Power demand	(kW)	22.5	30.5	30.5	42	42	42
Water demand	(m^3/h)	120-140	140-170	150-200	150-250	250-300	300-320
Water pressure	(MPa)	0.1	0.1	0.1	0.1	0.1	0.1
Weight of classifier with a mobile base	(kg)	ca. 15000	ca. 18000	ca. 19000	ca. 25500	ca. 21500	ca. 21900
Weight of classifier	(kg)	ca. 7500	ca. 9150	ca. 9950	ca. 11500	ca. 12600	ca. 13000

3. Use of a pulsating classifier for obtaining aggregates from natural resources

A prototype of the classifier was built and started its operation in 2002 in the "Dębówko" Mine of Aggregates, belonging to the Szczecińskie Kopalnie Surowców Mineralnych S.A.

The experience from the prototype tests was the basis for its implementation in various industrial plants. KOMAG classifiers were implemented in:

- KSM Sp. z o.o. in Borzęcin, belonging to CEMEX Polska (figure 2a),
- Zakład Produkcji Kruszyw i Prefabrykatów in Suwałki, owned by PPMD KRUSZBET S.A. (figure 2b),
- Gravel Pit PRInż. Surowce Sp. z o.o. in Januszkowice, then taken over by CEMEX Polska (figure 3a),
- Zakład Produkcji Betonów in Zdzieszowice, belonging to Przedsiębiorstwo Usługowo Handlowe "M " Sp. z o.o. in Kędzierzyn Koźle (figure 3b),
- Bierawa Gravel Pit, owned by CEMEX Polska (figure 4a),
- Rokitno Gravel Pit near Lublin (figure 4b).

Figure 2. K-100 pulsating classifiers in Borzęcin (a) and in Suwałki (b) [9,16].

Figure 3. K-100 classifiers in Januszkowice (a) and in Zdzieszowice (b) [9,16].

Figure 4. K-150 classifier in Bierawa (a) and K80 pulsating classifier in Lublin (b) [9,16].

Pulsating classifiers are used to extract from aggregates impurities in a wide spectrum of density: organic (particles of plants and wood, coal grains) and minerals (chalk, carbonates).

High efficiency of organic impurities separation from aggregates, exceeding 90% [19], enables to use a pulsating classifier for this process with a guarantee of obtaining satisfactory results, consistent with the user's requirements.

In the Polish aggregate manufacture plants, foreign design jigs are also used. There is, however, no available literature containing information on their effectiveness.

In order to increase the efficiency of pulsating classifiers at the KOMAG Institute of Mining Technology, a modernization and research and testing infrastructure is carried out, which enables their development [19,20-22].

An experience, gained in the result of subsequent implementations of pulsating classifiers, has shown that their efficiency in purifying aggregates depends on a number of factors.

The factors that have a significant impact on the efficiency of aggregate purification are the parameters of the material being processed. The most important of them include a type of impurities present in the raw material and the extent of the material graining.

Table 2 shows the results of laboratory tests on the impact of sand grains on the efficiency of aggregate purification.

Table 2. Specification of the results of density separation efficiency of gravel and gravel-and-sand feed [21,23].

Test		1	2	3	4	5	6
Share 3-0.5 mm in feed, (%)		0.0	0.0	25.0	25.0	50.0	50.0
Capacity, (t/h)		2.1	4.2	2.1	4.2	2.1	4.2
Unit load, (t/h m^2)		12	24	12	24	12	24
Separation efficiency%	< 1.5 (g/cm^3)	99.6	100.0	98.6	90.5	81.7	74.0
	1.5-1.8 (g/cm^3)	100.0	100.0	98.2	89.3	80.3	74.1
	2.1 (g/cm^3)	87.1	79.1	52.0	52.3	25.7	31.0

The tests confirmed that the efficiency of the jig aggregate beneficiation is reduced as the load of feed, the density of impurities and the participation of sand grains in the feed increase.

Unfavorable results of the distribution of high density particles of 2.1 g/cm^3 indicate that the purification process should be carried out in narrow grain classes and in the case of the benefication process the feed should not contain sand grains [21,23].

4. Use of a pulsating classifier for a recovery of minerals from mine heaps

A market demand and inquiries regarding a possibility of using a pulsating classifier for processing waste from mine heaps inclined for a development of a modernized version of the device at the KOMAG Institute of Mining Technology - classifier type K-102, which is shown in figure 5.

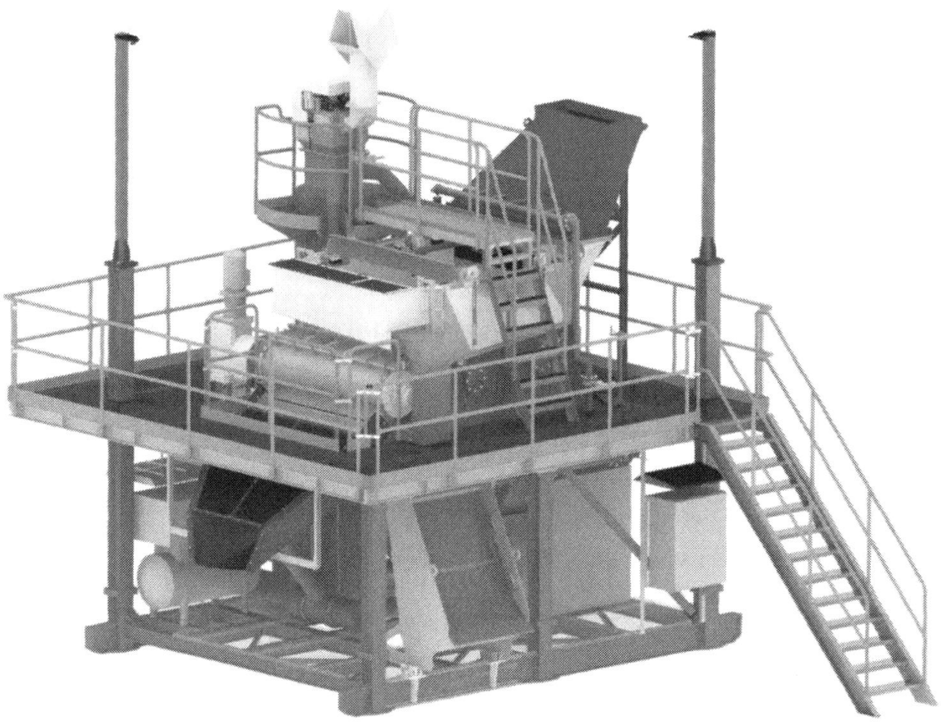

Figure 5. K-102 pulsating classifier on a mobile base for waste beneficiation from mine heaps [12,18].

In the result of hard coal production, a number of mine waste heaps have been generated, which cause many threats. These include fires, a release of gases that pollute the atmosphere, as well as leaching out of dangerous substances from heaps.

Due to a low efficiency of beneficiation processes, especially up to the mid- 20th century, a lot of coal grains were deposited on the heaps apart from the gangue [13,14,18].

In order to check the recovery efficiency of useful materials from waste heaps and for an adaptation of the K-102 classifier solutions, laboratory tests were carried out at the experimental stand [12,18].

The tests were performed on samples of waste materials from various mine heaps, which served as feed for the laboratory jig and during the technological test they were divided into two products: "light" (coal) and "heavy" (aggregate).

The samples of waste material in the 30-4 mm grain size class, taken from one of the mine waste heaps in Upper Silesia, were tested. The test results are summarized in table 3.

Table 3. Parameters of feed and beneficiation products (mine waste heap in Upper Silesia) – laboratory tests [12,18].

Fraction density, (g/cm^3)	Feed	"Light" product	"Heavy" product
	share, (%)	share, (%)	share, (%)
<1.5	5.63	76.30	0.38
1.5-1.8	5.70	16.84	4.87
>1.8	88.67	6.86	94.75
Sum	100.00	100.00	100.00
Output, (%)	100.00	6.92	93.08
Ash content, (%)	77.46	18.82	81.89
Calorific value, (MJ/kg)	4.13	25.68	2.52

In the result of the material beneficiation process, a concentrate product with an ash content of 18.82% and a calorific value of 25.68 MJ / kg as well as a 6.92% output was obtained. The waste product with 93.08% output was characterized by a small share of concentrate grains - 0.4% and an insignificant share of middlings grains - 4.9%. The ash content of the waste product was 81.89%, and its calorific value was 2.52 MJ/kg [12,18].

The test results showed that use of the gravitational beneficiation method in the K-102 pulsating classifier enables to obtain two full-value products.

A high density product, containing traces of organic matter, can be an alternative aggregate with a wide range of applications. The second product with a high calorific value and low ash content is steam coal concentrate.

The first implementation of the new K-102 classifier for a beneficiation of mine waste in the 35-3 mm grain class took place in 2015 at one of the mine heaps located in Upper Silesia.

In 2016, the second K-102 classifier was implemented. It was equipped with a state-of-the-art KOMAG control system, which allows a selection of operational parameters of the device depending on the required quantitative and qualitative parameters of the beneficiation products. Another pulsating classifier K-102 No. 3 was installed in 2017 [13,14,15].

The conducted research results show a high efficiency of the classifier, which enables to obtain beneficiation products of the quality parameters required by the user [15].

The K-102 pulsating classifiers in operation are shown in figure 6.

Figure 6. K-102 pulsating classifiers in the installations for a recultivation of mine heaps [15].

During an operation of the pulsating classifier, additional laboratory analyses of concentrate product samples were carried out, which showed very good quality parameters of this product. Selected results of tests, in operational conditions, are shown in table 4.

Table 4. Parameters of the concentrate product of a pulsating classifier [11,20].

No.	Total moisture, (%)	Ash content, (%)	Calorific value, (kJ/kg)
1	8.1	12.0	26138
2	7.8	11.3	26791
3	8.3	11.7	26651
4	9.2	10.8	26460
5	8.9	11.7	26609
6	10.8	10.3	26109

The processes of recovery of raw minerals from the mine heaps are usually based on the separation technology in cyclones with a heavy suspension fluid. Due to different technology, it is not possible to directly compare the effectiveness of the recovery processes. The technology based on heavy suspension fluid is a technology that allows for higher separation efficiency. However, it requires the use of an expensive magnetite particles. In addition, classifying in an liquid medium (water pulsating classifier) is a "pure" technology, unlike the classification in a heavy liquid due to the possibility of transferring the part of magnetite particles to the environment along with the separation products.

5. Design development of a pulsating classifier

A pulsating classifier, over the years, has been improved continually. Basing on the experience gained during the commissioning of subsequent K-100 classifiers, the K-150 classifier with increased capacity was constructed.

This capacity was obtained in the result of a reconstruction of the product's collection system - a more efficient rotary receiver with a larger capacity was used and an inclination of the collection channel was increased. In addition, the fines fraction collection system was changed by replacing four small conical nozzles with two large ones. A more efficient pulsation valve and a more powerful blower were also used [17].

Due to a need for an adaptation of the device to lower capacity, in 2012 the K-60 classifier was developed (figure 7).

In 2010 many changes were introduced in the construction of classifiers. The modernized version was named K-101. Among others a pulsation valve (figure 8) was changed, in which 4 adjustable plates were used, which allowed for an independent control of the air supply system. An auxiliary air vent was built in, enabling an additional adjustment of the working air pressure. A more effective silencer of the working air outlet was used [10].

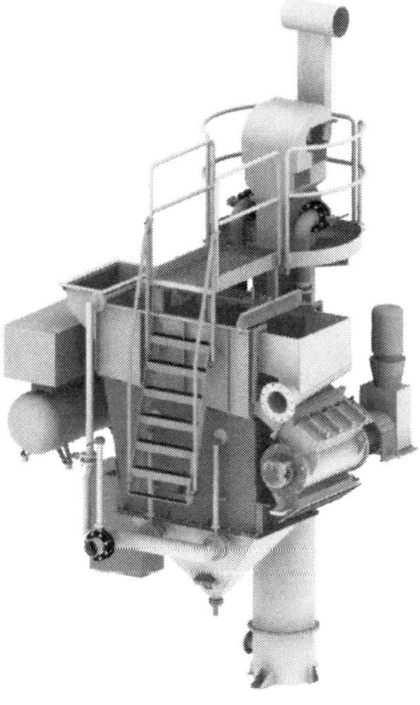

Figure 7. K-60 pulsating classifier [17].

Figure 8. ZP-4 pulsatory valve [10].

An alternative solution of the purified aggregate collection system was developed - an elastic receiver (figure 9). These changes made it possible to control an amount of the product received and to increase the maximum grain size in the feed up to 32 mm [10].

A new algorithm for a classifier operation was introduced, which enables a selection of the classifier operating modes for different types of feeds.

In addition, water and air dampers were installed to increase the range of control parameters, which make it possible to achieve higher aggregate purifying efficiency.

IOP Conf. Series: Materials Science and Engineering **545** (2019) 012006 doi:10.1088/1757-899X/545/1/012006

In the K-102 classifier installed in 2016, the KOMAG classifier control system was used for the first time (figure 10). It allowed to integrate a control of the entire beneficiation node and to obtain a number of benefits in terms of, among others:

- an optimization of the capacity of devices in the beneficiation node,
- a reduction of electricity consumption,
- an increase of service life of the discharge system under variable load conditions,
- an increase of the scope of process monitoring and a generation of a process documentation,
- a reduction of downtime,
- an access to data from one place [15].

Figure 9. Classifier elastic receiver [10].

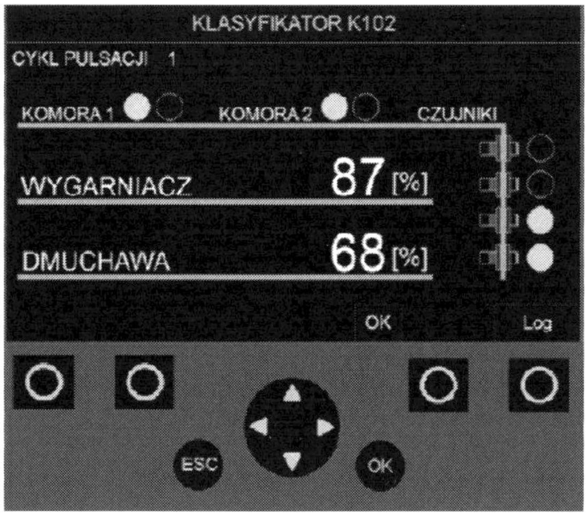

Figure 10. Main screen of K-102 No. 2 pulsating classifier control panel [15].

The classifier was also equipped with a pulsation pressure measurement system (figure 11), improving a control capabilities of the classifier and enabling to obtain higher efficiency and more favourable beneficiation indicators [15].

Figure 11. Classifier No. 3 - pressure sensor in air chambers of operational compartment [15].

6. Summary

Many years of experience of the KOMAG Institute of Mining Technology in the construction and technological selection of jigs for a beneficiation of hard coal allowed for a development of a new device - a pulsating classifier, designed for a separation and purification of aggregates [2, 9].

KOMAG pulsating classifiers confirmed their advantages characterized by a reliable operation and low operational costs, at high efficiency of separating impurities from the aggregate.

The next step at the KOMAG development of equipment for a beneficiation of minerals was a use of a modernized version of the pulsating classifier for a beneficiation of mine waste.

The results of industrial tests of the mine waste beneficiation confirmed a high efficiency of the device, enabling to obtain both steam coal concentrate and so-called alternative aggregate.

The previous implementations of the pulsating classifiers showed that they are machines with a wide range of applications, and among their many advantages an ability to adapt the design to user's requirements in terms of construction conditions, expected capacity and quality of commercial products should be mentioned.

In order to increase the classifier efficiency, at the KOMAG Institute of Mining Technology, a continuous research and development work is carried out, using a laboratory jig stand. This work, through changes in technological and construction solutions, makes it possible to increase efficiency of minerals' beneficiation processes.

References

[1] Blaschke W 2009 *Przeróbka węgla kamiennego – wzbogacanie grawitacyjne (* IGSMiE PAN – Wydawnictwo, Kraków)

[2] Matusiak P and Kowol D 2013 Maszyny do przeróbki mechanicznej konstruowane w ITG KOMAG *Maszyny Górnicze* **2** pp 71-76

[3] Kowol D, Jędo A, Osoba M and Łagódka M 2002 Badania efektywności wzbogacania miału węglowego o uziarnieniu 20(30)0.5 mm w wodnych osadzarkach pulsacyjnych typu KOMAG. *KOMEKO Nowoczesne technologie i systemy mechanizacyjne do przeróbki surowców mineralnych* Szczyrk pp 113-124

[4] Mijał W and Tora B 2018 Development of dry coal gravity separation techniques *IOP Conf Series: Materials Science and Engineering* 427

[5] Sampaio C H, Aliaga W, Pacheco E T and Petter E 2008 Coal beneficiation of Candiota mine by dry jigging *Fuel Processing technology* **89**(2) 198-202

[6] Sanders G J, Ziaja D and Kottmann J 2002 Cost efficient beneficiation of coal by ROMJIGs and BATAC jigs *Coal Preparation* **22**(4) pp 181-197

[7] Bahrami A, Ghorbani Y, Mirmohammadi M, Sheykhi B and Kazemi F 2018 The beneficiation of tailing of coal preparation plant by heavy media cyclone *Int J Coal Sci Technol* **5**(3) pp 374-384

[8] Zhang L and Xia X 2014 A Model Predictive Control for Coal Beneficiation Dense Medium Cyclones *IFAC Proceedings* **47**(3) pp 9810-9815

[9] Lenartowicz M, Matusiak P, Kowol D and Łagódka M 2010 Innowacyjne rozwiązanie klasyfikatora pulsacyjnego do oczyszczania surowców mineralnych *Maszyny Górnicze* **3-4** pp 119-125

[10] Matusiak P and Kowol D 2013 Możliwości poprawy jakości kruszywa poprzez zastosowanie klasyfikatora pulsacyjnego typu KOMAG *Prace Naukowe Instytutu Górnictwa Politechniki Wrocławskiej* **136** Studia i Materiały **43** pp 109-118

[11] Różański Z, Suponik T, Matusiak P, Kowol D, Szpyrka J, Mazurek M and Wrona P 2016 Coal recovery from a coal waste dump *E3S Web of Conferences* **8** 01052 92016

[12] Matusiak P and Kowol D 2016 Zastosowanie osadzarkowego procesu wzbogacania do odzysku surowca ze składowiska odpadów górniczych *Min. Sci., Miner. Aggreg.* **1** pp 115-125

[13] Matusiak P and Kowol D 2017 Technologia odzysku koncentratu węglowego z odpadów pogórniczych poprzez wdrożenia urządzeń typu KOMAG *Maszyny Górnicze* 2017 **2** pp 42-52

[14] Matusiak P and Kowol D 2017 Wdrożenia urządzeń typu KOMAG do odzysku surowców z odpadów pogórniczych *Materiały na konferencję: IV Polski Kongres Górniczy Kraków*, pp 986-995

[15] Matusiak P and Kowol D 2018 Zastosowanie inteligentnych rozwiązań w procesach produkcji kruszyw mineralnych *Kruszywa mineralne* **2** pp 133-142

[16] Matusiak P and Kowol D 2012 Klasyfikator pulsacyjny jako sprawdzone urządzenie do oczyszczania surowców mineralnych *Prace Naukowe Instytutu Górnictwa Politechniki Wrocławskie, Studia i Materiały, Gór. Geol.* **134/41** pp 191-199

[17] Kowol D and Matusiak P 2014 Zastosowania klasyfikatora pulsacyjnego typu KOMAG do oczyszczania trudnowzbogacalnych surowców mineralnych *Mining Science - Mineral Aggregates* **21(I)** pp 99-108

[18] Kowol D and Matusiak P 2016 Możliwości zastosowania klasyfikatora pulsacyjnego do rewitalizacji składowisk odpadów kopalnianych *Innowacyjne rozwiązania rewitalizacji terenów zdegradowanych* **8** pp 163-172

[19] Osoba M 2007 Osadzarki wodne pulsacyjne KOMAG do przeróbki żwiru i piasku *Prace Naukowe Instytutu Górnictwa Politechniki Wrocławskiej* **119**(48) pp 127-136

[20] Not published materials ITG KOMAG, 2002-2018

[21] Kowol D and Łagódka M 2012 Badania wpływu udziału ziaren piaskowych na skuteczność procesu osadzarkowego wzbogacania kruszywa *Maszyny Górnicze* **4** pp 54-59

[22] Kowol D and Matusiak P 2015 Badania skuteczności osadzarkowego oczyszczania kruszywa z ziaren węglanowych. *Min. Sci., Miner. Aggreg.* **1** pp 83-92

[23] Matusiak P, Kowol D and Nieckarz R 2012 Nowe rozwiązania klasyfikatora pulsacyjnego do oczyszczania surowców mineralnych *Maszyny Górnicze* **4** pp 49-53

IOP Conf. Series: Materials Science and Engineering **545** (2019) 012007 doi:10.1088/1757-899X/545/1/012007

Integration of a selected district heating system with an installation for cogasification of coal and biomass

T Iluk A Sobolewski M Szul and T Billig

Institute for Chemical Processing of Coal, Zabrze, Poland

tiluk@ichpw.pl

Abstract. The article presents a topic related to generation of heat and power in small and medium size coal power units which are able to comply with the emission benchmark of 550 g of CO_2/kWh through utilization of biomass. The concept for coal and biomass cogasification installation is based on GazEla reactor, which has been developed by Institute for Chemical Processing of Coal. Authors also present a discussion on issues related to technological scheme of the gasification unit which has been integrated with a selected district heating system. The system has been set up for generation of hot technological water and electrical power. Focal point of the installation is a fixed bed gasification reactor. For the system to provide means enabling cogeneration of heat and power, downstream from the generator a high-temperature gas dedusting system, a unit for recovery of heat from process gas coupled with separation of tar contaminants as well as a gas engine were proposed. A sorted chart presenting historical data for annual demand for heat together with the most important assumptions used during the analyses are discussed. Finally authors present results of calculations for the integrated unit that are further broadened by sensitivity analyses for change of gasification reactor's efficiency, as well as engine's efficiency for heat and power generation.

1. Introduction

In recent days there is a clear trend for shift in legal and formal regulations regarding energy sector to impose the need for search of new and innovative solutions that will allow for generation of heat and power with higher efficiencies, lower environmental emissions as well as will utilize higher amounts of renewable energy source (RES), i.a of biomass [1, 2].

The subject is particularly important for currently operating small and medium size of power generating systems which operate on traditional primary fuels, eg. coal barley, which face today the challenge of adaptation to new regulations, eg. environmental ones. Another important aspect of the problem is also connected with emission of CO_2, which mainly pertain to the planned UE directive which will impose the criteria for CO_2 emission below the level of 550 g/kWh. Power stations with higher emissions of CO_2 will be obligated to pay additional money for their emissions within the frame of ETS scheme. This on the other hand will influence highly the final cost of energy production. This is particularly importation in the light of recent rise in the prices of CO_2 emission equivalent cost, which at the beginning of 2019 reached the level of 23.7 €/t, and which according to long term forecasts will stay on a rise for many years to come [3, 4].

One of the solutions discussed frequently for small and medium scale units, which allows for an efficient utilization of heat and power, both regarding environmental and energetic considerations is gasification technology [5, 6]. This technology enables generation of heat and power in cogeneration units based on piston engines. The solid fuels through gasification are converted into syngas which after upgrading and cleaning has excellent properties as a gaseous fuel. Combustion of such gas in piston engines provides energy for operation of electricity generators and produces high temperature flue gases, which through cooling and heat recovery allow for generation of useful heat. This heat can be than connected with another source of high temperature heat, which in gasification installations is found from cooling of the process gas [6, 7].

The article presents topics related to integration of a gasification installation with a selected local heating grid, which allows for exchange of coal fired water boiler operated for production of hot water during the entire year, as well as utilization of the generated power when considering the 550 g/kWh criteria, through utilization of coal and biomass cogasification technology.

2. Gasification technology – GazEla reactor

The technology for gasification of solid fuels is based on reactors, which depending of efficiency as well as the final means of process gas utilization can be divided into fixed bed reactors, fluidised bed reactors and entrained bed reactors. Process gas, generated within them, may be utilized for direct combustion and generation of high temperature flue gases, feeding of cogeneration units based on piston engines or gas turbines, as well as production of synthetic gases (eg. CH_4 and H_2) or liquid fuels and chemical substances such as methanol or DME.

Institute for Chemical Processing of Coal for many years develops gasification technologies for i.a coals, biomasses and wastes. The research and development works are carried out mainly for energetic utilization of process gases in cogeneration units as well as for production of synthetic gaseous and liquid fuels.

For sake of the analyses presented hereinafter, one of the technologies developed by IChPW has been chosen. The gasification process considered in this article is carried out in fixed bed GazEla reactor. The reactor was designed and developed for the in view of generation of process gas from solid fuels i.e. mainly biomass, for cogeneration of heat and power in gas piston engines. This technology is dedicated for small and medium scale energy systems eg. district heating stations as well as heat and power stations, companies that have technological lines operated on hot water or technological steam as well as companies with large internal demand for electricity.

The gasification installation consists of four main technological blocks, ie. fuel preparation and feeding, GazEla fixed bed gasification reactor, gas cleaning unit and cogeneration unit based on gas piston engine.

Central point of the system is the GazEla reactor, which is a vertical, cylindrical reactor operated under fixed bed conditions. In the vertical axis of the reactor a central pipe is fitted which allows for recovery of process gas directly from gasification zone. Fuel is fed into the reactor from top, while gasification agent is fed into three zones: under reactor's grate, into its middle part as well as above the level of fuel bed. Proper selection of type and stream of gasifying agent is crucial for regulation of load of each process zones of the reactor. This mechanism provides means for control of temperature profiles observed within the reactor and obtaining right process conditions in relation to physicochemical parameters of gasified feedstock [8, 9, 10]. Figure 1 presents a technological scheme of the reactor with denoted specific process zones.

IOP Conf. Series: Materials Science and Engineering **545** (2019) 012007 doi:10.1088/1757-899X/545/1/012007

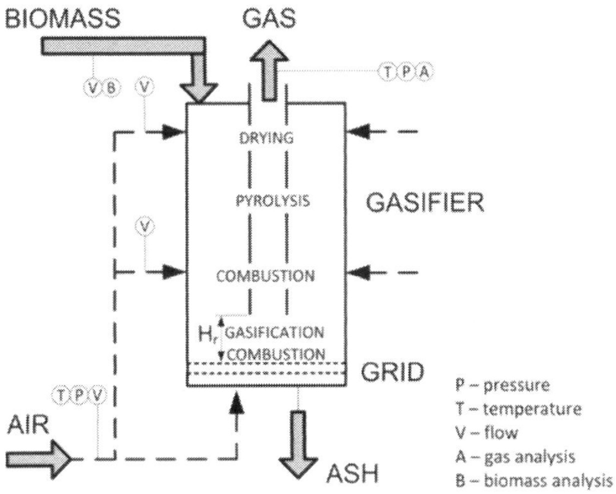

Fig. 1. Technological scheme of GazEla reactor with denoted specific process zones.

In Figure 2 pictures of a pilot GazEla reactor, which is located in Clean Coal Technology Centre of IChPW are presented.

Fig. 2. Pictures of pilot GazEla fixed bed reactor (top and bottom part of the reactor)

3. A reference unit of a coal and biomass cogasification installation integrated with a selected district heating system

As a reference point for the analysis, a heating system located in Miejski Zakład Energetyki Cieplnej w Świdnicy Sp. z o.o. has been taken [11].

Analysis has been done in order to assess the possibility to exchange one of the boilers for a coal and biomass cogasification installation, which allows for yearlong generation of hot water according to current production of the boiler as well as production of electricity while exercising the 550g CO_2/kWh rule.

Figure 3 presented below, represents a sorted chart for annual demand for heat for the selected boiler in year 2018. A minimal daily average outside temperature (atmosphere) was equal to $T_{z\ min}$= -12.0°C, whilst a maximal daily average outside temperatures was equal to $T_{z\ max}$= 29.0°.

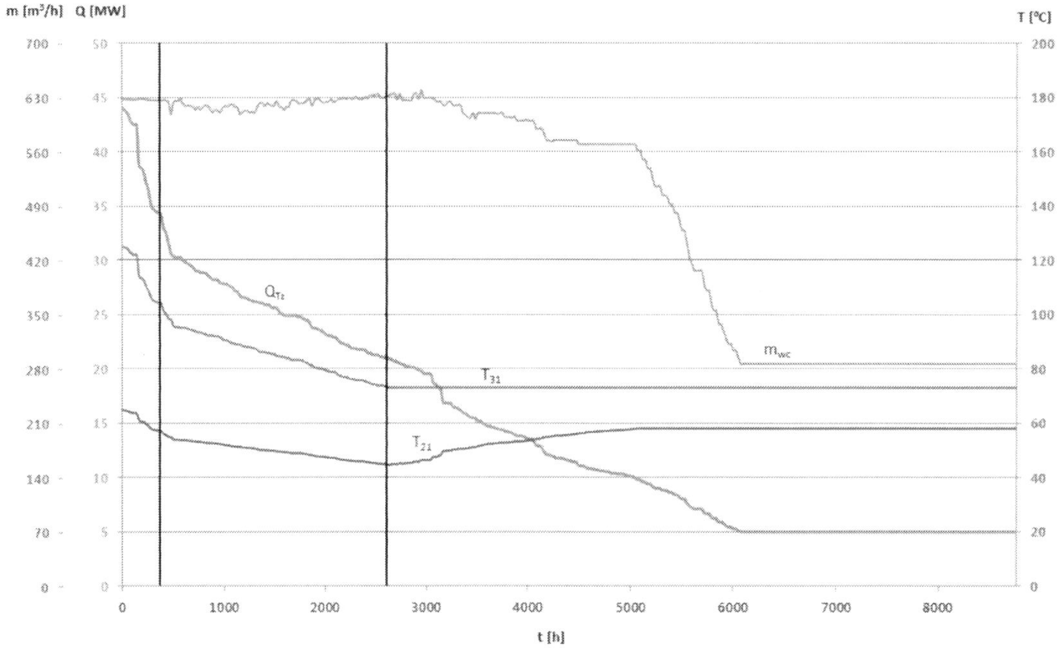

Fig. 3. Sorted annual heat demand chart for the selected water boiler.

Year average daily outside temperatures was equal to $T_{z\,avg}$= 11.2°C. Accordingly, heat demand of the boiler was equal to $Q_{Tz\,min}$= 44.0 MW and $Q_{Tz\,max}$= 5.0 MW. Parameters of the heating water at inlet and outlet from the boiler for $T_{z\,min}$ and $T_{z\,max}$ are presented in Table 1. Line T_{21} represents temperature of inlet (incoming) hot water, whilst line T_{31} depicts temperature of outlet (heated) water. Line m_{wc} describes stream of heating water.

Table 1. Characteristic parameters for selected district heating system.

Parameter	Unit	Value
Minimal power of the heating station, Q_{Tzmax}	MW	5
Maximal power of the heating station, Q_{Tzmin}	MW	44
lowest daily-average outside temp., $T_{z\,min}$	°C	-12
highest daily-average outside temp., $T_{z\,max}$	°C	29
minimal temperature of inlet water, T_{21min}	°C	45
maximal temperature of inlet water, T_{21max}	°C	65
minimal temperature of outlet water, T_{31min}	°C	73
maximal temperature of outlet water, T_{31max}	°C	125
minimal stream of water, m_{wcmin}	m^3/h	286.4
maximal stream of water, m_{wcmax}	m^3/h	639.5

A model based on the concept for gasification unit used for energy and heat production in piston engine has been taken for analysis. Figure 4 shows a block diagram of biomass and coal gasification unit integrated with the selected district heating unit (from Fig. 3).

IOP Conf. Series: Materials Science and Engineering **545** (2019) 012007 doi:10.1088/1757-899X/545/1/012007

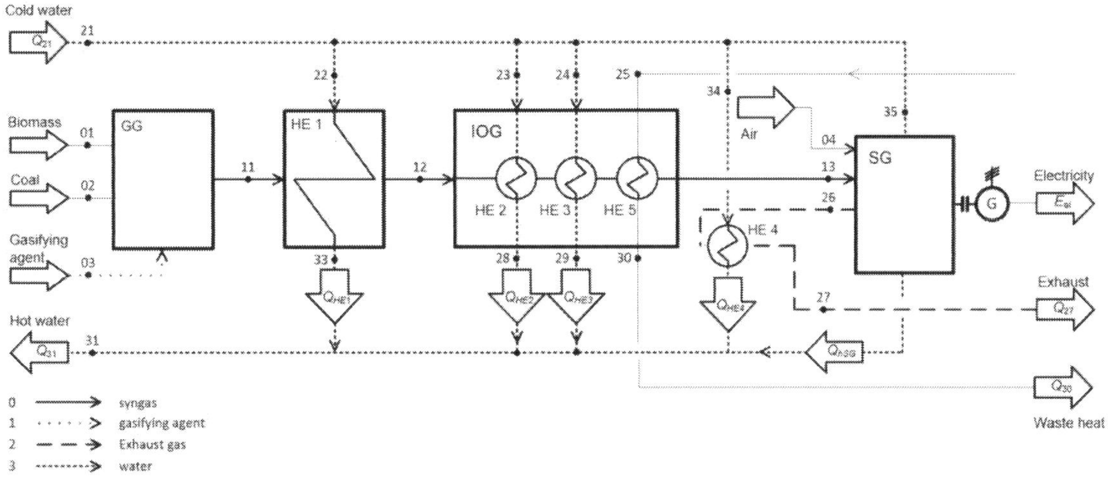

Fig. 4. Block diagram of biomass and coal gasification installation for heat and power production for demand of a district heating unit.

The system presented on Fig. 4 comprises of a gas generator (GG), integrated with a dedusting unit – adiabatic process, as well as gas cooling unit ((HE 1). Next element of the installation is a gas cleaning block (IOG) which separates tar contaminants and is based on three heat exchangers (HE 2, HE 3, HE 5). Cleaned and cooled process gas is directed into piston engine (SG), where heat Q_{hSG} and power E_{el} are cogenerated. Additionally fuel gases are further cooled in heat exchanged (HE 4), where heat Q_{HE4} is generated. Heat streams Q_{HE1}, Q_{HE2}, Q_{HE3} obtained through process gas cooling, afterwards are connected with heat streams generated in piston engine Q_{hSG} and Q_{HE4} and finally together comprise the total usable heat, necessary for heating purposes. In heat exchanger HE 5 the final lowering of process gas temperature to 40°C is done according to technical demands of piston engines. Due to low temperature of water obtained in HE 5, the heat stream Q_{30} is not used for heating of usable hot water and thus is treated as waste.

Table 2 presented bellow shows physicochemical properties of coal and biomass used in the analysis.

Table 2. Physicochemical properties of coal and biomass

Denotation	**Unit**	**Coal**	**Biomass**
Moisture content	%	19.1	22.1
Ash content	%	11.8	1.0
Volatile matter content	%	25.5	63.7
Carbon content	%	56.2	40.3
Hydrogen content	%	3.1	4.5
Nitrogen content	%	0.8	0.1
Oxygen content	%	8.0	32.0
Sulfur content	%	1.0	0.0
Lower heating value	J/g	20 693	14 279

Due to the low power of the installation, it has been assumed that for gasification a fixed bed reactor will be used, which for gasification of biomass is characterized by cold gas efficiency equal to ca. 63.0%. The efficiency is defined by the following equation:

$$\eta_{GG} = \frac{(\dot{m}\cdot LHV)_{11}}{(\dot{m}\cdot LHV)_{01}+(\dot{m}\cdot LHV)_{02}} \tag{1}$$

Where:
m_{11} – mass stream of process gas,
LHV_{11} – lower heating value of the process gas,
m_{02} – mass stream of coal,
LHV_{02} – lower heating value of coal,
m_{01} – mass stream of biomass,
LHV_{01} – lower heating value of biomass,

For purpose of the analysis it has been assumed that air will be the gasification agent.

In order to assure parameters of the heating water for the selected district heating unit during the whole year, as well as to fulfil the CO_2 emission regulation of 550 g of CO_2/kWh, it is necessary to feed the reactor with stream of fuel equal to 2 313.6 kg/h. Mass fractions of biomass and coal in the blend are equal to 71.7% of biomass and 28.3% of coal. Stream of air directed to the gasifier is equal to 3559.9 kg/h. Assumed heating value of the process gas (dry) equals to 4.91 MJ/Nm^3.

Hot process gas recovered from the reactor has 750°C and is cooled in heat exchangers. Stream of heat obtained from gas in the exchangers has been determined with use of the following formula:

$$\dot{Q}_{HEi} = \dot{m}_{gas} \cdot \eta_{HEi} \int_{T_{outi}}^{T_{ini}} c_p(T)dT \quad (2)$$

Where:
Q_{HEi} – stream of heat recovered in heat exchanger,
T_{ini} – gas temperature at the inlet,
T_{outi} – gas temperature at the outlet,
$c_p(T)$ – specific heat capacity of gas at function of temperature,
It has been assumed that efficiency of HEs is equal to 99%.
Total useful heat of the unit is calculated with use of the following formula:

$$\dot{Q}_{heat} = \dot{Q}_{hSG} + \sum_{i=1}^{4} \dot{Q}_{HEi}\ [kW] \quad (3)$$

Another assumption is that HE 1 cools process gas from 750°C to 500°C. This gas is further directed to dedusting. Afterwards, in HE 2 gas is further cooled to temperature 110°C. In HE 3 and HE 5 the gas is cooled to 40°C. Here condensation of water vapour takes place. After the final step of cooling, the gas is directed to combustion in piston engine. Flue gases from the engine are cooled in HE 4 to temperature of 130°C.

For calculations it was set that technological water is the heating medium. Parameters of the water were determined based on the annual demand for technological hot water.

Parameters of the water are accordingly, at summer equal to 73/58°C, while at winter 125/65°C. The installation reaches maximal heating power in the intermediate term, when parameters of water drop down to 74/45°C. For the assumed temperature of heating water, it is not possible to cool down the process gas to the level of 40°C in HE 3. For this reason the additional heat exchanger HE 5 has been set and it is cooled by water chilled by a draft fan cooler. It has been assumed that pinch point in the coolers is equal to 5K. Figure 5 presents a sorted chart of annual temperatures of technological water and their relevant power demands for HE 3 and HE 5. For technological water with parameters above 105/57.5°C it was determined that all heat will be recovered in heat exchanger HE 5.

IOP Conf. Series: Materials Science and Engineering **545** (2019) 012007 doi:10.1088/1757-899X/545/1/012007

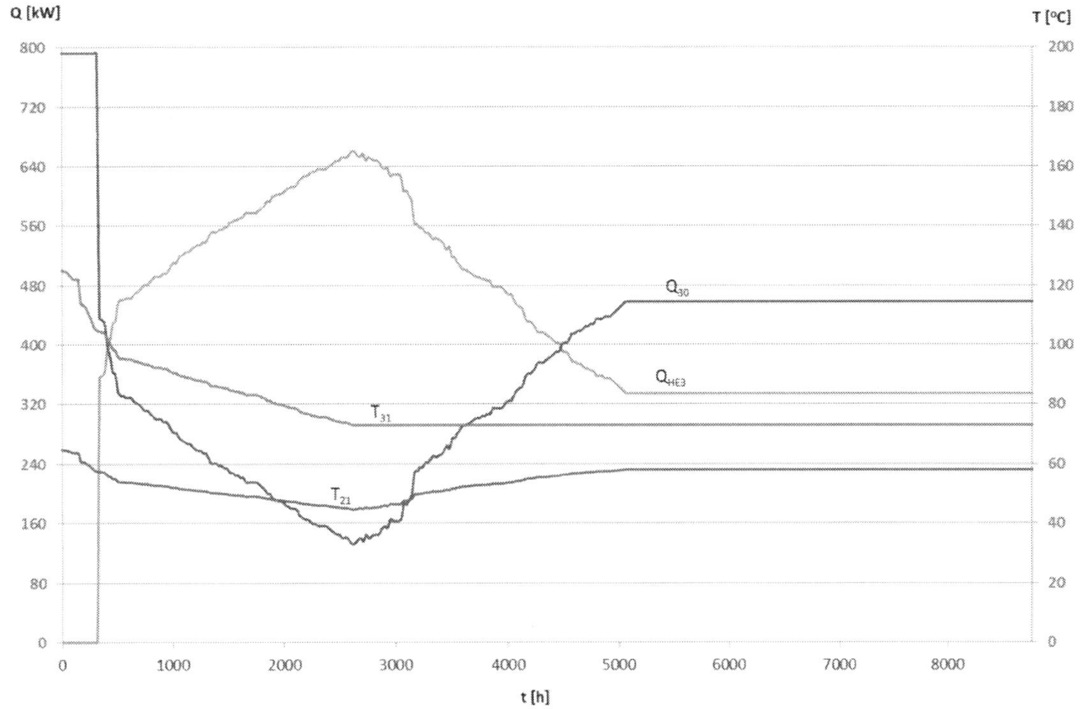

Fig. 5. Sorted chart of annual temperatures of technological water with depicted powers of the heat exchangers HE 3 and HE 5.

A cumulated heat production was calculated by the following formula:

$$Q_{heatref} = Q_{hSG} + \sum_{i=1}^{4} Q_{HEi} \left[\frac{GJ}{year}\right] \quad (4)$$

while heat recovered by each heat exchanger was calculated with:

$$Q_{HEi} = \int_{t_0}^{t_{max}} \dot{Q}_{HEi}(t)dt \left[\frac{GJ}{year}\right] \quad (5)$$

$$Q_{heat} - Q_{heatref} \cdot A \quad (6)$$

where A, is availability of the installation, set as a ratio of the amount of operational hour during a year:

$$A = \frac{t_{work}}{8760} \quad (7)$$

It was assumed that during a year the installation will work 7500 h.

A cumulated heat recovered by the subsequent steps of heat exchangers is presented in table 3.

Table 3. Cumulated heat recovered by heat exchangers

Water temperatures	**°C**	**73/58-73/45**	**73/45-105/57.5**	**> 105/57.5**
Work time	h	6 120	2 328	312
Heat generated				
Heat exchanger 1, Q_{HE1}	GJ	10 379.6	3 948.3	529.2
Heat exchanger 2, Q_{HE2}	GJ	14 898.3	5 667.2	759.5
Heat exchanger 3, Q_{HE3}	GJ	7 430.4	3 954.7	0
Heat exchanger 4, Q_{HE4}	GJ	2 308.8	878.3	117.7
Gas engine, Q_{hSG}	GJ	60 421.1	22 983.7	3 080.3
Waste heat, Q_{30}	GJ	7 514.8	1 730.3	761.9
Net electricity, E_{elnet}	kW	2 195.6	2 200.9	2 182.4

IOP Conf. Series: Materials Science and Engineering **545** (2019) 012007 doi:10.1088/1757-899X/545/1/012007

Cooled gas is than burned in piston engine. Efficiency of power generation of the motor is defined by:

$$\eta_{elSG} = \frac{\dot{E}_{elgross}}{(\dot{m} \cdot LHV)_{13}} \quad (8)$$

Heat generation efficiency of the engine was defined as:

$$\eta_{hSG} = \frac{\dot{Q}_{hSG}}{(\dot{m} \cdot LHV)_{13}} \quad (9)$$

It was assumed that efficiency of heat and power generation in a piston engine fuelled by process gas is accordingly equal to η_{hSG} = 49.2% and η_{hSG} = 37.7%. Thus efficiency of cogeneration in the engine is equal to:

$$\eta_{SG} = \frac{\dot{Q}_{hSG} + \dot{E}_{elgross}}{(\dot{m} \cdot LHV)_{13}} \quad (10)$$

The calculations take into account also the power consumed for internal purposes of the unit. It was assumed that the index of internal demand α is equal to 10%. Additionally it was set that power demand to the draft cooler E_{elCT}, is linearly demendant on its capacity which is variable in time. Generated net electrical power was calculated with:

$$\dot{E}_{elnet} = (1 - \alpha)\dot{E}_{elgross} - \dot{E}_{elCT} \quad (11)$$

Total efficiency of cogeneration is determined as a sum of efficiencies of heat and power generation from a fuel:

$$\eta_{CHP} = \frac{\dot{E}_{elnet} + \dot{Q}_{heat}}{(\dot{m} \cdot LHV)_{01} + (\dot{m} \cdot LHV)_{02}} \quad (12)$$

Table 4 presents all the most important assumptions and results of the performed balance calculations.

Table 4. Collation of assumptions and results of calculation

Parameter	Unit	Value
Assumed work time	h/year	7 500
Assumed own need indicator	%	10.0
Fuel mixture stream	kg/h	2 313.6
Gasifier thermal power	kW	10 334.1
Biomass content in fuel	%	71.7
Cold gasification efficiency	%	63.0
Gas engine electricity production efficiency	%	37.7
Gas engine heat production efficiency	%	49.2
Combined heat and power efficiency (for summer parameters)	%	69.63
Calculated total average combined heat and power efficiency	%	70.48
Heat exchangers heating power for water at temperatures 73/58°C	kW	5 000.0
Heat exchangers heating power for water at temperatures 73/45°C	kW	5 318.4
Heat exchangers heating power for water at temperatures 125/65°C	kW	4 665.6
Gross electricity production	kW	2 454.4
Net electricity production for water at temperatures 73/45°C	kW	2 206.9
Net electricity production for water at temperatures 125/65°C	kW	2 182.4
Total demand for chemical energy in fuel	GJ/year	279 070
From coal	GJ/year	101 158
From biomass	GJ/year	177 912
Total heat production	GJ/year	137 357
Total waste heat in draft fan cooler	GJ/year	10 007
Total net electricity produced	MWh	16 473.9

Basing on the abovementioned calculations, a sensitivity analysis has been done for a set of parameters changing based on efficiency of the unit. Fig. 6 presents the impact of change of cold gas efficiency of the gasifier as well as efficiency of power generation and heat generation of the engine, under the assumption that rest of the process parameters remain constant. For nominal efficiency of gasification in the reactor (63.0%) a total efficiency of the installation is equal to 69.63%. For the change of value in range of -3% to +3%, the overall efficiency of the installation changes in the range of 67.13 % ÷ 72.12%. While, for the change in efficiency of power generation of the engine from -3% to +3%, the change of overall efficiency of the installation changes in the range of 67.74% ÷ 71.52%.

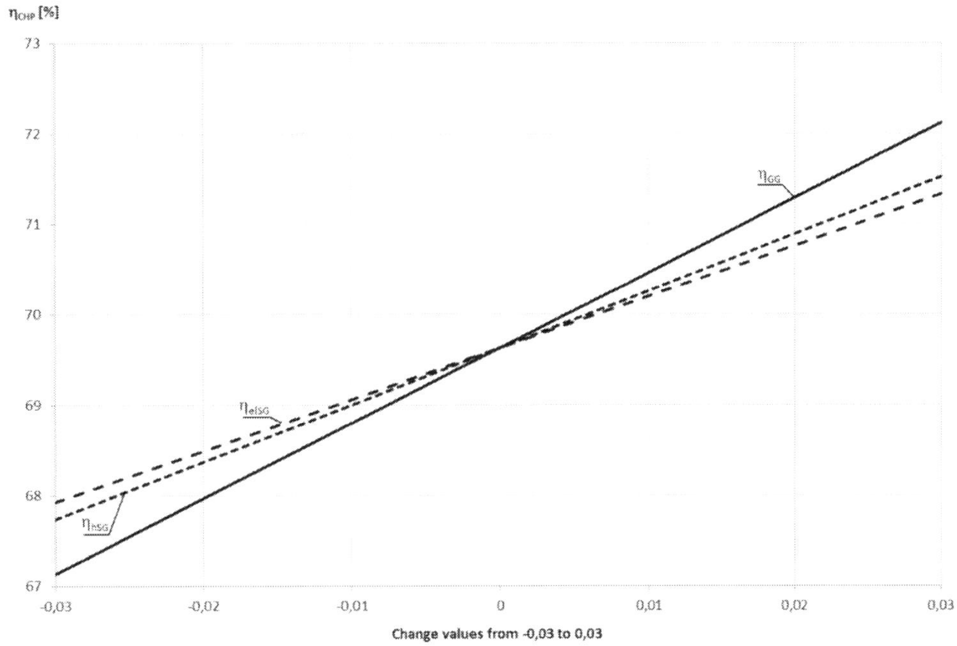

Fig. 6. Influence of change of gasifier's efficiency as well as heat and power generation efficiency of the engine for the overall efficiency of the installation.

4. Conclusion

Development of energy technologies is directed towards development of innovative, ecological solutions that enable highly efficient production of energy carriers. This problem concerns both large scale energy systems as well as medium and small scale distributed units.

Currently, for the range of small and medium power size units, one of the most intensively developed technologies that enable highly efficient generation of heat and power, both from energetic and environmental points of view is gasification of solid fuels. This technology enables production of a combustible gas from solid fuels, i.a. coal, biomass, which after cleaning can be effectively combusted in highly efficient cogeneration units based on piston engines. This provides opportunity for generation of power as well as heat from process of cooling down of high temperature flue gases from the engine, units of its cooling as well as high temperature process gas during its cleaning.

The article presents the issue related to integration of a gasification installation with a selected district heating installation with changing demand for power, ie. from 5 MW to 44 MW. In order to guarantee parameters of the heating water for a selected unit annually, as well as to fulfil the benchmark of CO_2 emission under the level of 550 g/kWh, it is necessary to feed the gasifier with 2 313.6 kg/h of fuel. Mass fraction of biomass and coal in the blend is equal to 71.7% for biomass and 28.3% for coal. Under assumed efficiency of gasification reactor 63.0% respectively efficiency of power and heat generation of the engine η_{elSG} = 37.7% and η_{hSG} = 49.2%, the overall efficiency of the installation is equal to 69.63%.

Acknowledegements
The results presented in this paper were obtained during the research project entitled „Doskonalenie technologii zgazowania biomasy oraz odpadów w reaktorach ze złożem stałym i ze złożem fluidalnym (GazEla, IZPS, IZOP)." (IChPW No 11.19.018), financed by the Ministry of Science and Higher Education.

References

[1] Skorek-Osikowska A Bartela Ł and Kotowicz J 2017 *App. Ener.* **200** p 73-88 DOI: 10.1016/j.apenergy.2017.05.055.
[2] Bartela Ł Kotowicz J Remiorz L Skorek-Osikowska A and Dubiel K 2017 *Ren. Ener.* **112** p 425-443
[3] https://handel-emisjami-co2.cire.pl/st,34,514,me,0,0,0,0,0,ceny-uprawnien-do-emisji-co2.html
[4] https://wysokienapiecie.pl/12584-ceny-uprawnien-emisji-co2-wyniosa-25-euro
[5] Skorek-Osikowska A Kotowicz J Uchman W 2017 *Energy* 2017 **141** 1764-1778 DOI: 10.1016/j.energy.2017.11.106
[6] Skorek J Kalina J *Gazowe układy kogeneracyjne* Wydawnictwa Naukowo-Techniczne, Warszawa 2005
[7] Kotowicz J Sobolewski A Iluk T 2013 *Energy* 2013 **52** p 265-278
[8] Kotowicz J Iluk T 2011 *Rynek Energii* **94** p 47-52
[9] Sobolewski A Kotowicz J Iluk T Matuszek K *Przem. Chem.* **6** p 794-799.
[10] Billig P Ściążko M Sobolewski A *Zgazowarka ze złożem stałym* Patent PL-208616, 2010
[11] http://mzec.swidnica.pl/

IOP Conf. Series: Materials Science and Engineering **545** (2019) 012008 doi:10.1088/1757-899X/545/1/012008

State-of-The-Art Pulsating Jigs of KOMAG Type

P Matusiak and D Kowol

KOMAG Institute of Mining Technology, Division of Preparation Systems,
37 Pszczyńska Street, 44-101 Gliwice, Poland

pmatusiak@komag.eu, dkowol@komag.eu

Abstract. For several decades the KOMAG Institute of Mining Technology has been involved in a construction of pulsating jigs - devices for a gravity beneficiation of hard coal. The conducted research and conceptual work enables their constant development, resulting in their high reliability and efficiency. The article presents the latest design solutions and examples of implementations of the KOMAG state-of-the-art pulsating jigs in the Polish hard coal mines. The operational principle and possibilities of an innovative control system of the KOGASTER SSWO jig node are also discussed

1. Introduction

A beneficiation has a decisive influence on the quantitative and qualitative parameters of commercial products. It is the basic process in Polish preparation plants of mineral raw materials. The beneficiation process of hard coal in jigs is an extremely important part of a production process of commercial assortments, serving the purpose of improving the quality of extracted raw material and achieving quality parameters of the products required by recipients. The quality of raw coal is improved by subjecting it to a series of technological processes, among others a beneficiation process in pulsating jigs [1]. The following paragraphs present a historical outline and solutions of the latest devices designed at the KOMAG Institute of Mining Technology for a beneficiation process of hard coal. Particular attention is paid to state-of-the-art solutions of OM fine grain jigs and OS medium grain as well as to new solutions used in them.

2. Historical background of KOMAG pulsating jigs

The first jig constructions were developed at KOMAG in 1955. Since then, pulsating jigs in the Polish preparation plants have been the basic devices, used in hard coal beneficiation processes.

Initially, the devices, having air chambers placed along the axis of the jig, next to the working bed [2], were constructed. They had the following markings:

- OBM (fine coal air pulsating jig),
- ODM (fine coal double trough air pulsating jig) ,
- OBZ (grain coal air pulsating jig).

These jigs were characterized by big dimensions and big weight.

In the 1970s, at Komag, the first Polish jigs were designed with air chambers located under the sieve deck, built in lower boxes, transversely to the longitudinal axis of the jig [3]. This solution is used until today. Jigs for beneficiation processes of various grain classes are constructed, including:

- fine grain jigs - OM (figures 1 and 2), grain class 20-0(0.5) mm,

IOP Conf. Series: Materials Science and Engineering **545** (2019) 012008 doi:10.1088/1757-899X/545/1/012008

- medium grain jigs - OS, grain class 80 (50) -0 (2) mm,
- grain jigs - OZ, grain class 120 (200) -20 mm.

A construction is subject to continuous modernizations, in line with the expectations of customers, both in terms of the shape geometry of the of lower boxes, fixing of sieve plates, an improvement of automatic reception control devices of heavy products, pulsation valves and working air collectors [4].

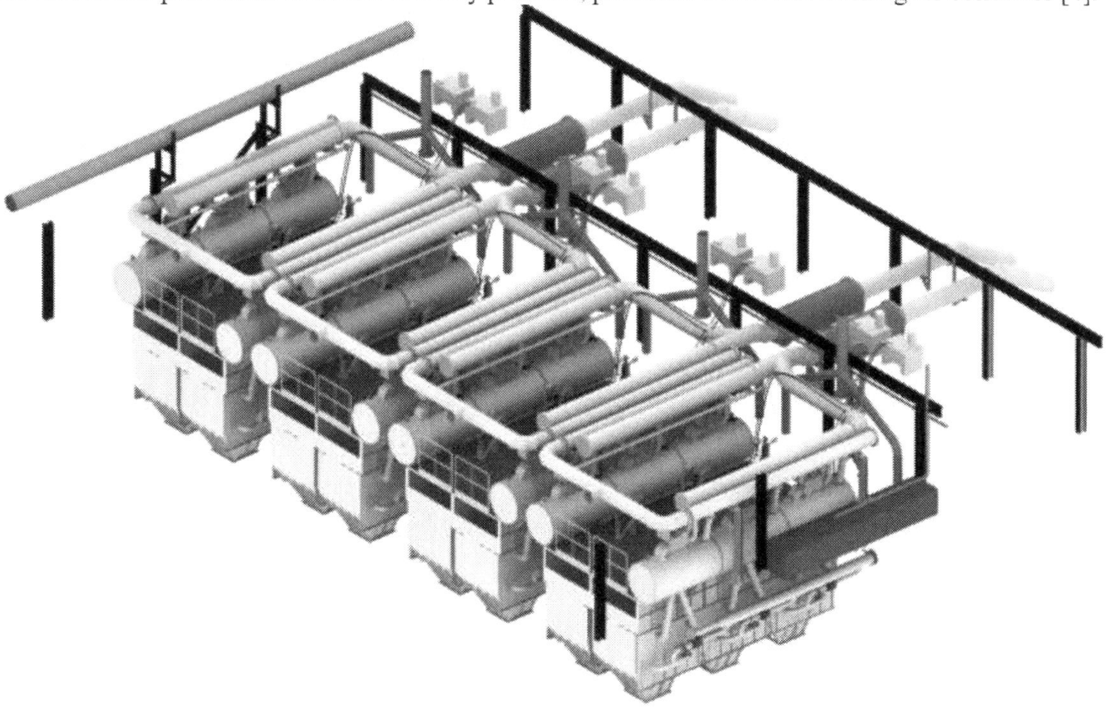

Figure 1. "Pniówek" Mine – a replacement of old OM24D3E double trough jigs for OM15 single trough.

3. Construction and operational principle of a pulsating jig

3.1 Construction of main components

The jig is composed of working compartments, which include lower boxes and upper boxes called working trough. The jig upper boxes are equipped with sieve plates, working air collectors, pulsating air timing gear and collecting devices of beneficiation products. In the lower boxes there are two pulsating chambers in the shape of a bell, to which working air is supplied, causing a pulsating movement of water and the collectors feeding the lower water at adjustable flow [5].

3.2 Jig principle of operation

A beneficiation process in jigs consists in using the rate of sinking in the water of grains differing in density. The beneficiated material undergoes cyclical loosening in the pulsating water medium, which causes its stratification and transport to the surface of the sieve bed towards the separation zone of the beneficiation products. The pre-prepared mixture of hard coal grains, fed to the working box, is subject to pulsation in the water medium on the sieve plates in the subsequent working compartments of the jig [6]. The pulsating movement of water, in the working chamber, generated by compressed air, causes stratification of the material [6]. During the water upward movement, the material resting on the sieve, is raised, then it falls down towards the sieve deck.

IOP Conf. Series: Materials Science and Engineering **545** (2019) 012008 doi:10.1088/1757-899X/545/1/012008

Subjecting the material to multiple pulsations (50-80 cycles per minute) results in a separation of the material into layers, from the grains with the highest falling speed, located on the sieve board, to the grains with the lowest falling speed, carried to the surface of the material under beneficiation. The stratified material moves continuously towards the separation zone, at the end of the jig working compartment [3].

On the basis of measurement signals and control of the collection system of beneficiation products, the beneficiated material is separated into the upper product (coal concentrate) which flows over the overflow threshold, ending the single jig compartment and the bottom product (waste), discharged to bucket conveyors through the collecting channel located below the sieve deck [5, 7].

3.3 Designs of KOMAG jigs

Table 1 [8, 9] presents a list of pulsating jigs designed at KOMAG in the years 1955-2019.

Table 1. List of jigs designed at KOMAG over the years 1955-2019 [8, 9].

Production period	Jigs type	Place of application
1955-1970	OBM12, OBM15, OBSZ15, OBZ10, OBZ12	Coal mines: "Anna", "Dębieńsko", "Knurów", "1-Maja", "Nowy Wirek", "Mysłowice", "Rydułtowy", export to China, India and Vietnam
1971-1985	ODM10, ODM18, ODZ15, OM12, OM12-2, OM12-3, OM12-3S, OM12G3, OM12P3, OM12L3, OM18P3, OM18L3, OM24-3, OM24D, OM24B3, OM24D3, OZ18L, OZ12, OZ12L, OZ12P3, OS36D3, OZ36D3, OC8, OC10	Coal mines: "Bogdanka" "Borynia", "Dębieńsko", "Gliwice", "Halemba", "Knurów", "Krupiński", "Jankowice", "Makoszowy", "Marcel", "Pniówek", "Dymitrow", "Rydułtowy", "Sośnica", "Staszic", "Wujek", "Szczygłowice", "Wawel", "Zabrze", "Zofiówka", "Moszczenica", export to Brazil, India and Romania.
1986-2014	OM8L2e, OM8L2E, OM10L2E, OM15P3E, OM12P3E, OMPE-3x6,5, OM18L3E, OM18P3E, OM20P3E, OM20L3E, OM24P4E, OM24L4E, OM24D3E, OS24D3E, OM30-3E, OM18 3x8, OM30D3E, OS30D3E, OM30, OZ18, OZ18L3E, KOD Jig	Coal mines: "Andaluzja", "Anna", "Barbara-Chorzów", "Bogdanka", "Borynia", "Budryk", "Dębieńsko", "Halemba", "Jastrzębie", "Jasmos" "Knurów", "Krupiński", "Marcel", "Pniówek", "Rozbark", "Rymer", "Rydułtowy-Anna", "Sośnica", "Staszic", "Szczygłowice", "Wawel", "Wujek", "Zofiówka", export to Czech Republic and India. „Budryk" Mine for destoning of the run-of-mine.
2003-2018	Pulsating Classifiers K-100, K-150, K-50, K-80, K-101, K-102, K-151	KSM sp. z o.o. in Borzęcin, PPMD KRUSZBET S.A in Suwałki, PRInż. Surowce Sp. z o.o. in Januszkowice, PUHM "M+" Sp. z o.o. in Kędzierzyn Koźle, Żwirownia Bierawa, PRESTO Emil Potręć, Rokitno, Rent-Pol - Przezchlebie
2014-2020	OM30, OS18 L, OS18P, OM15L, OM15P, OS4, OM20, OS18L OS18P, OM24, OM15L and OM15P, OM30, OM24	Coal mines: "Sośnica", "Krupiński", "Pniówek", "Budryk", "Zofiówka", ZG "Eko-Plus", ZG "Sobieski". "Bielszowice" Mine, "Bogdanka" Mine.

Within 65 years, KOMAG developed technical documentations of over 200 types of jigs and 120 other beneficiation devices.

4. New solutions used in KOMAG jigs

The followings examples show implementations of jigs, designed at KOMAG over the years 2016-2020.

4.1 Coal Mechanical Processing Plant at "Pniówek" Mine

In 2016, at Komag, a modernization project of a beneficiation node for the Coal Mechanical Processing Plant at "Pniówek" Mine was developed. A modernization mainly involved a replacement of 4 OM24D3E double trough jigs for 8 OM15 single trough jigs, and a use of more efficient bucket conveyors. The first system of the upgraded beneficiation node was started in 2017. The next ones will start in 2019 and 2020.The modernized beneficiation node (figure 3) consists of 4 beneficiation systems, each of them is equipped with:

- two fine coal jigs OM15L and OM15P (20-0 mm),
- two bucket conveyors B-1000 of capacity 250 t/h and 150 t/h.

Figure 2. OM30 fine coal jig „Sośnica" Mine [8].

Figure 3. Coal fines beneficiation node at "Pniówek" Mine after modernization.

Designed OM15 fine grain jigs working in the 20-0 mm grain size class (figure 6) achieve a capacity up to 300 t/h. They use a new solution of the Festo pneumatic actuator control system (figure 4), reducing the sensitivity of the system to the control air contamination. A modernization of the plant was carried out in cooperation with the Wrębowa Company. The jig uses a new type of product discharge (fiugre 5), extending the working surface of the working deck and reducing disturbances of the material layer at the threshold.

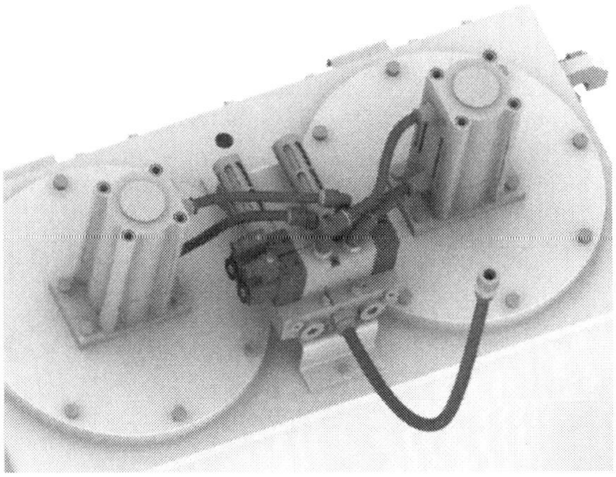

Figure 4. KOMAG pulsation valve.

Figure 5. Discharge of product - sliding damper.

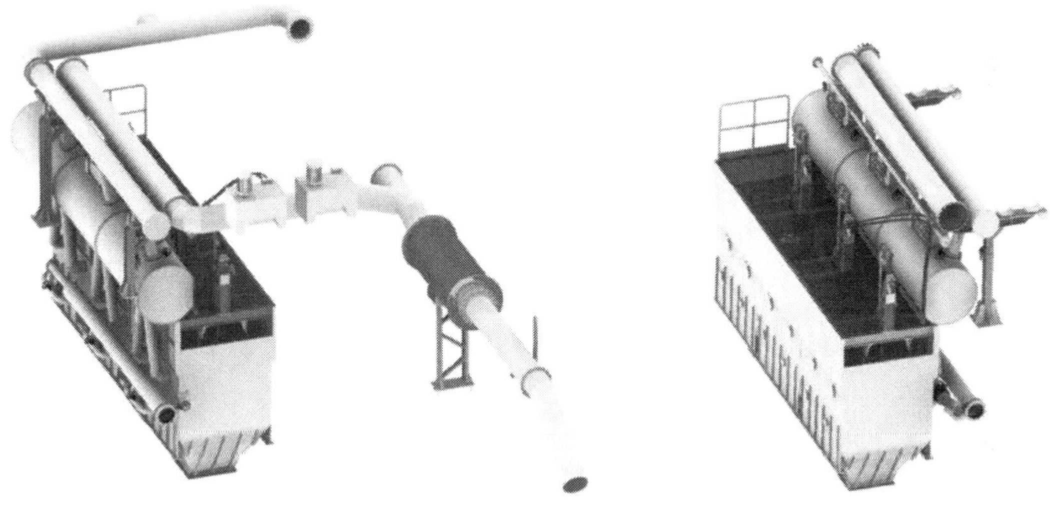

Figure 6. OM15L and OM15P new jigs - "Pniówek" Mine.

4.2 Coal Mechanical Processing Plant at "Budryk" Mine

In 2017, new jigs and bucket conveyors were designed at KOMAG for the Coal Mechanical Processing Plant at the "Budryk" Mine. The first system of the upgraded beneficiation node is scheduled for commissioning in 2019. The modernized beneficiation node (figure 7) consists of 2 beneficiation systems, each of which is equipped with:

- 3 medium grain jigs OS18 (L and P) (70-2 mm),
- 5 bucket conveyors B-1000 of capacity 220t/h.

OS18 medium grain jigs (figure 8), of 250 t/h capacity, are intended for a beneficiation of grain sizes 80-0.5 mm. Slotted sieves S = 5mm are used in them. Additionally the hypertrophic product is directed to the UP 1000x1500 crushers, where it is recycled up to 12 mm after crushing. The second beneficiation system is equipped with:

- 2 fine coal jigs OM20 (12-0 mm),
- 4 bucket conveyors B-1000 of capacity 220 t/h.

IOP Conf. Series: Materials Science and Engineering **545** (2019) 012008 doi:10.1088/1757-899X/545/1/012008

A construction of the secondary beneficiation system enables an operation of only one jig. In the secondary fine coal jigs OM20 (figure 9), for a beneficiation in the grain size class of 12-0 mm with a maximum capacity of up to 250 t / h, a rubber sieve deck is used. A modernization was carried out in cooperation with the following companies: Famur S.A., Pemug S.A. and Fugor Sp. z o.o.

Figure 7. Two beneficiation systems - OS18 medium grain jigs.

Figure 8. The secondary beneficiation system – OM20 two fine coal jigs.

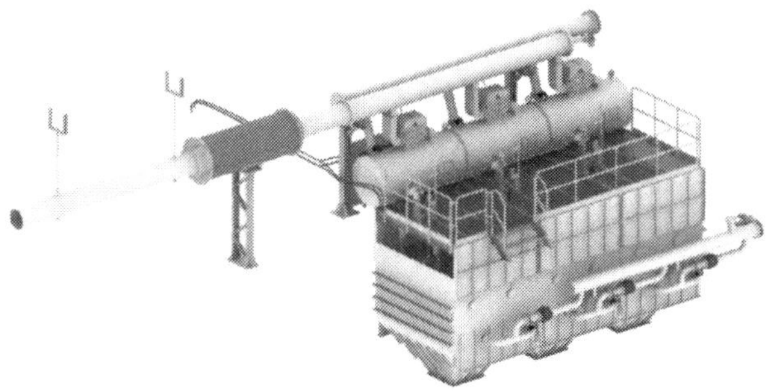

Figure 9. OM20 fine coal jig - "Budryk" Mine.

4.3 Coal Mechanical Preparation Plant at "Borynia-Zofiówka-Jastrzębie" Mine," Zofiówka" Mine

In 2018, a jig design was made for the Coal Mechanical Processing Plant at "Borynia-Zofiówka-Jastrzębie" Mine, "Zofiówka" Mine. The new system will be started in 2019.The first modernized beneficiation node system (figure 10) consists of a OM24 fine grain jig (20-2 mm), of maximum capacity up to 640 t/h. It uses perforated sieves decks Lv 25x5 30x20. The presented jig is the only device in Poland having the working bed width of 3500 mm.

Figure 10. OM24 fine coal jig - "Zofiówka" Mine.

In the modernized beneficiation system an innovative solution of the jig feed boxes, with an overflow channel and a spray, which improve an uniformity of the material distribution before the jig and a better use of the working bed (figure 11) was used. The modernization is under realization in cooperation with the following companies: Carbo-Eco Sp. o.o and Fugor Sp. z o.o.

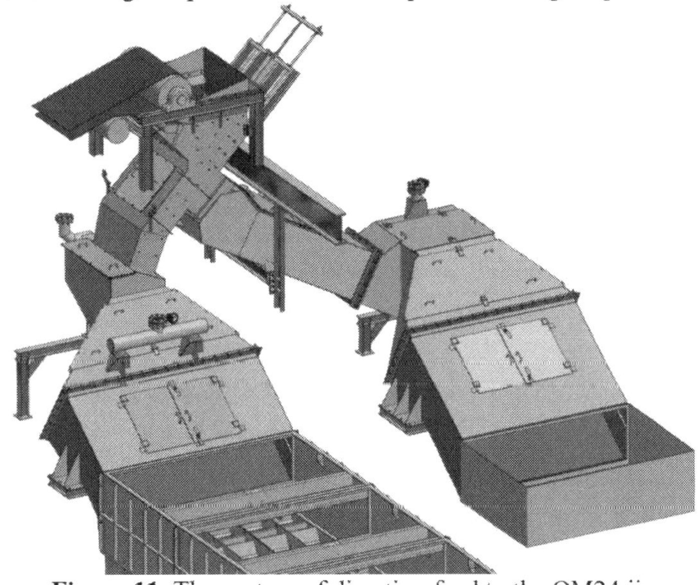

Figure 11. The system of directing feed to the OM24 jig.

4.4 Coal Mechanical Processing Plant at "Sobieski" Mining Plant

In 2019 the jigs for Coal Mechanical Processing Plant at "Sobieski" Mining Plant will be commissioned. The project was completed in 2018. The modernized beneficiation node consists of two fine coal jigs (figure 12) OM15L and OM15P working in the grain class of 30-2 mm, with a maximum capacity of up to 320 t/h each.

Figure 12. OM15L and OM15P fine coal jigs at "Sobieski" Mining Plant.

They will replace worked out two-compartment Allmieral jigs. The new jigs will beneficiate three products. An installation of three working compartments replacing two jigs, due to an insufficient length of the working area, required shortening of the working compartments to 2200 mm. The decks made of slotted sieves S = 5 mm, were used in them. Modern WZL1 2.8x6.0 screening machines, washing the material below 2mm, shown in (figure 13), were applied in front of the jigs.

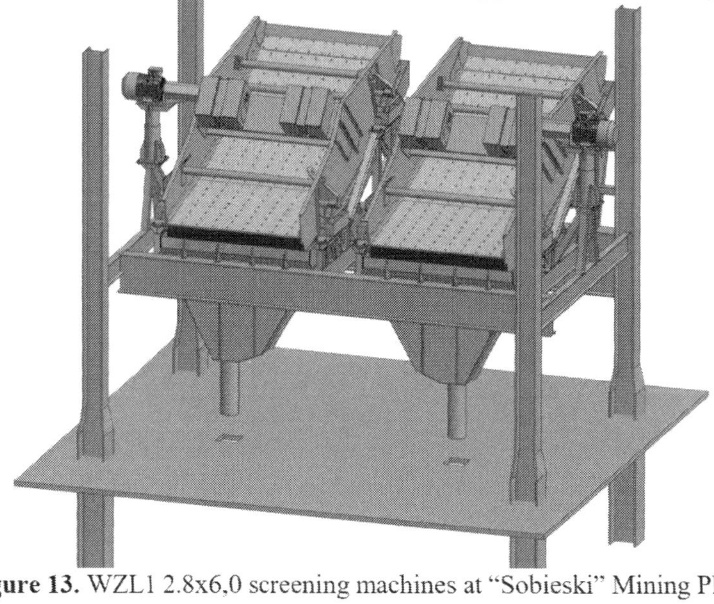

Figure 13. WZL1 2.8x6,0 screening machines at "Sobieski" Mining Plant.

5. Summary

Modernized beneficiation node systems in the "Budryk", "Borynia-Zofiówka-Jastrzębie", "Zofiówka" Mines and "Sobieski" Mining Plant will be equipped with KOGASTER SSWO - modern jig node control system (figure 14).

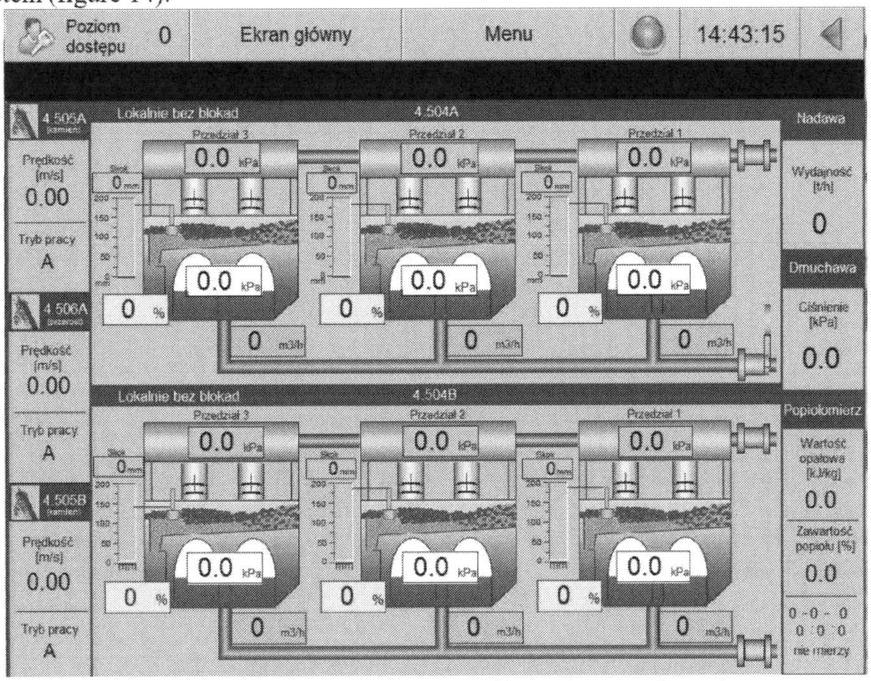

Figure 14. KOGASTER SSWO jig node control panel.

Among many functions of the system, it enables, among others:

- a control of water pulsation process, with a possibility of changing the length and number of phases (inlet, outlet and gaps between them) of the pulsation cycle, independent for each working compartment,
- an automatic adjustment of heavy products collection including a measurement of the culvert,
- a selection of the opening range of heavy product discharges,
- monitoring of the jig operation with a possibility of signalling irregularities,
- a speed control of bucket conveyors,
- a measurement and recording of monitored parameters, individually for each compartment [10]

The system, based on online measurements, shows ash content and it estimates calorific value of produced coal concentrate.

6. Conclusions

The modernization projects on beneficiation systems, conducted at KOMAG, were carried out in cooperation with their users, which resulted in a creation of innovative technological and design solutions. KOMAG researchers also cooperated with other research and development centres, universities and producers. During more than 60 years of jig design history at KOMAG, all their components were subject to structural changes. The main innovative features of developed jigs are:

- an efficiency increase,
- an improvement of beneficiation accuracy indicators,
- a reduction of technological water consumption,
- a reduction in the installation area which resulted in a weight reduction.

An improvement of the coal concentrate quality, at given technological parameters, was achieved, among others, due to:

- an installation of KOGASTER jig control systems,
- an automation of media control,
- a modernization of the air supply system (pulsation valves),
- a modification of the heavy product collection assembly.

References

[1] Blaschke S 1982 *Przeróbka mechaniczna kopalin.* Wydawnictwo Śląsk. Katowice
[2] Matusiak P, Kowol D and Jędo A 2011 Rozwój konstrukcji i technologii w osadzarkach pulsacyjnych typu Komag do wzbogacania węgla kamiennego i innych surowców mineralnych *Innowacyjne i przyjazne dla środowiska techniki i technologie przeróbki surowców mineralnych Bezpieczeństwo - Jakość - Efektywność KOMEKO* pp73-86
[3] Osoba M and Jędo A 2000 Wpływ modernizacji osadzarek na uzyskiwane parametry jakościowe produktów wzbogacania *Produkcja surowców mineralnych z uwzględnieniem problemów ochrony środowiska* pp 1-7
[4] Matusiak P and Kowol D 2013 Maszyny do przeróbki mechanicznej konstruowane w ITG KOMAG *Masz. Gór.* **2** pp 71-76
[5] Cierpisz S and Kowol D 2012 Wpływ zmian parametrów nadawy na efekty działania pływakowego układu sterowania odbiorem produktów osadzarki *ITG KOMAG*
[6] Kowol D, Lenartowicz M, Łagódka M and Matusiak P 2009 Przegląd rozwiązań konstrukcyjnych zespołów rozrządu powietrza pulsacyjnego w osadzarkach pulsacyjnych *Automatyzacja Procesów Przeróbki Kopalin*
[7] Kowol D, Lenartowicz M and Łagódka M 2009 Rozwiązania konstrukcyjne układów odprowadzania produktu dolnego w osadzarkach pulsacyjnych *Innowacyjne i przyjazne dla środowiska systemy przeróbcze surowców mineralnych w aspekcie zrównoważonego rozwoju* pp 225-236
[8] Matusiak P and Kowol D 2018 Rozwój osadzarek pulsacyjnych typu KOMAG *Masz. Gór.* **2** pp 40-52
[9] Kowol D, Matusiak P and Łagódka M 2018 Możliwości zwiększenia efektywności procesu produkcji miałów węgli koksowych poprzez wzrost dokładności wtórnego wzbogacania półproduktu w osadzarce pulsacyjnej *Masz. Gór.* **2** pp 53-64
[10] Matusiak P, Kowol D and Łagódka M 2016 Nowe wdrożenia wzbogacalników pulsacyjnych typu KOMAG *Innowacyjne i przyjazne dla środowiska techniki i technologie przeróbki surowców mineralnych. Bezpieczeństwo - Jakość – Efektywność KOMEKO* pp 175-189

IOP Conf. Series: Materials Science and Engineering **545** (2019) 012009 doi:10.1088/1757-899X/545/1/012009

SCADA Class Software of the KOGA Control system of Jig Beneficiation Node

J Rogala-Rojek, K Stankiewicz and S Jendrysik

KOMAG Institute of Mining Technology, Division of Mechatronic Systems,
37 Pszczyńska, 44-101 Gliwice, Poland

jrogala@komag.eu, kstankiewicz@komag.eu, sjendrysik@komag.eu

Abstract. The article presents a proprietary visualization software for the coal beneficiation process, which enables monitoring of the jig node operation and full control of trends relating to all signals from sensors installed in the system and connected to the controller, among others a movement of the float, a position of the culvert and a position of the overflow threshold. The results of analyses regarding the selection of PID regulator settings, determined on the basis of industrial research, are also presented.

1. Introduction

Increasing the quality of products obtained in the process of beneficiation, carried out in jigs, in coal mechanical processing plants, requires a implementation of solutions from the broadly understood industrial automation. An indispensable component of these solutions is computer software, the application of which is related to the hardware layer, communication layer, control, analysis, acquisition, archiving and visualization of data. The jig node control system, developed at the KOMAG Institute of Mining Technology, provides: an operational monitoring and control of its correct functioning, an automatic emergency shutdown and remote manual or sequential stopping and starting of the jig and the machines associated with it.

2. Characteristics of the jig control system

An efficient progress of each technological process requires an implementation of a proper control system of each machine being a part of the jig node. In the process of hard coal beneficiation, taking place in pulsatory jigs, a very important factor, determining their proper operation, includes an efficient and precise control of raw material feeding and removal of the beneficiated product. It is connected with a necessity of a complex automation of the node, which can ensure a proper interaction of the jig with the feed system and bucket conveyors collecting products from the jig. An exemplary structure of the jig node control system is presented in figure 1 [1, 2].

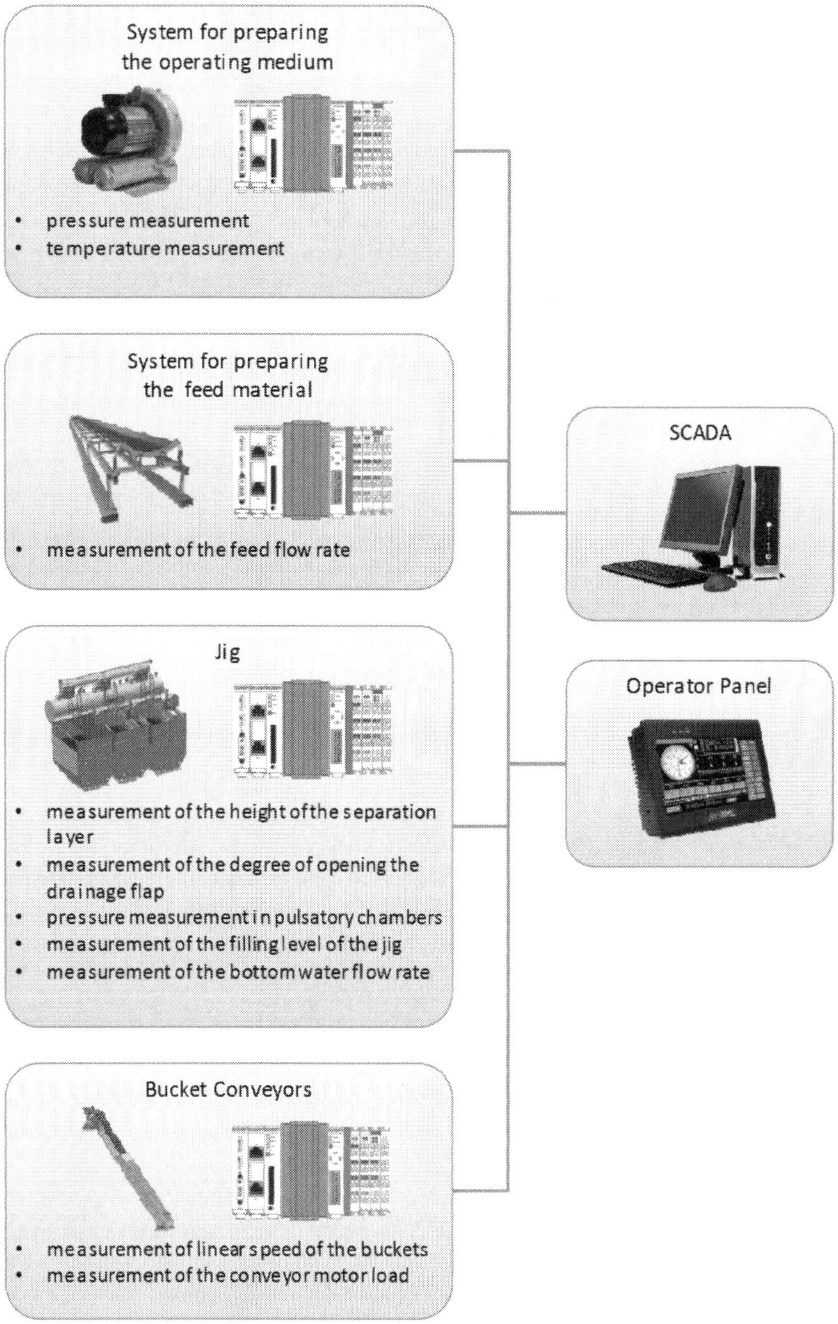

Figure 1. Structure of the distributed control system of jig beneficiation node [3, 4].

The KOMAG Institute of Mining Technology has been conducting research projects on a development of systems for a control of a jig node for several years. In 2010, the first prototype of the system was installed in the processing plant of the Budryk mine. Initially, the system was used to control a pulsatory jig, and then, in the first half of 2011, it was extended by an integration with the bucket conveyor control [1]. The KOGA jig node control system, developed at KOMAG, was also implemented in 2016 in Sośnica mine, branch of KW S.A. (Coal Company) [5].

The control system developed at KOMAG, built on the basis of a PLC programmable controller, provides monitoring and control of the correct operation of the jig node, automatic, emergency shutdown and remote, manual and sequential stopping and start-up of the jig and machines that cooperate with it [1].

The basic functions of the control system include [1]:

- a pulsation control, with a possibility a changing its parameters in relation to the length and number of cycles,
- an automatic adjustment of the heavy products collection,
- a loosening stabilization of the beneficiated material layer,
- a measurement and recording of quantities related to an operation of the machine,
- an automatic control of the jig in the place of its installation as well as a possibility of control through a dispatching system,
- an automatic and manual (from the operator's panel) speed control of bucket conveyors,
- an automatic unloading of the jig.

A very important element of the control system related to a production process is the so-called human-machine interface HMI. The role of such interfaces, in modern systems based on PLCs, is played by operator's panels and SCADA software (Supervisory Control And Data Acquisition) [2].

They provide a textual or graphical representation of phenomena occurring during a realization of a technological process. They also make it possible to track and control a flow of the process via a terminal, such as a computer monitor or an operator's panel screen. Use of these tools increases work comfort, both for the personnel directly operating the process, as well as the management staff, allowing to obtain information about the process in a quick and easy way and at the same time enabling as follows [2]:

- monitoring an operation of the machine:
  - switching the machine on/off,
  - giving orders to the operator concerning e.g.: the correct sequence of switching on and off of a given machine (recipe),
  - displaying process messages and information,
- process diagnostics, including:
  - monitoring process parameters,
  - scaling of values in specific engineering units,
  - displaying text messages, alarm/data history,
  - defining alarm thresholds,
  - signalling faults and alarm confirmations,
- setting values and changes of thresholds of regulation loops,
- conducting manual control of the object,
- keeping of archives of process data with an option of their printing,
- generating of a process visualization.

3. IT system supervising the technological process (SCADA)

An operation of the node is monitored by a visualization application (SCADA), whose main screen is shown in figure 2. The application is installed on a computer, in the control room of the processing plant. Its main functionality includes a visualization of the process and keeping archives of the measurement data, however, the range of its capabilities is much wider. By using an appropriate logging system, it allows trained users to access advanced control functions and settings of regulation loop parameters.

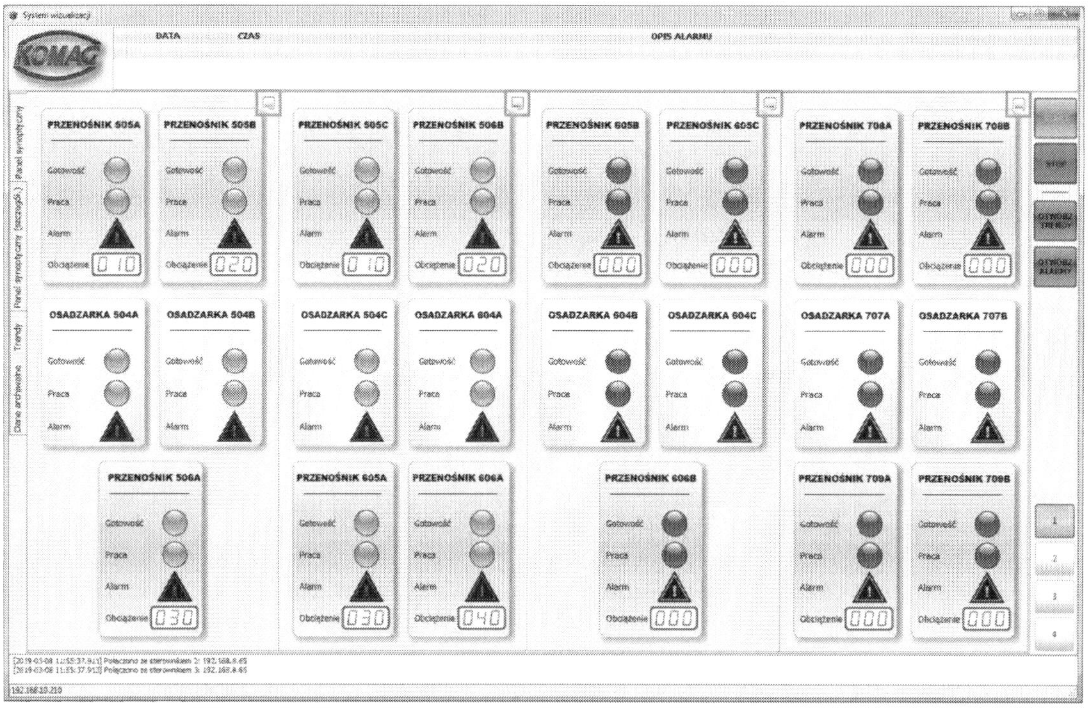

Figure 2. View of the main window of visualization application.

The main screen shows basic information about an operation of all the jigs and bucket conveyors installed in the processing plant. Individual machines are grouped according to their connections to the control cubicles. The software user (after clicking the button marked with a blue frame) is able to view more detailed information on an operation of individual machines (figure 3).

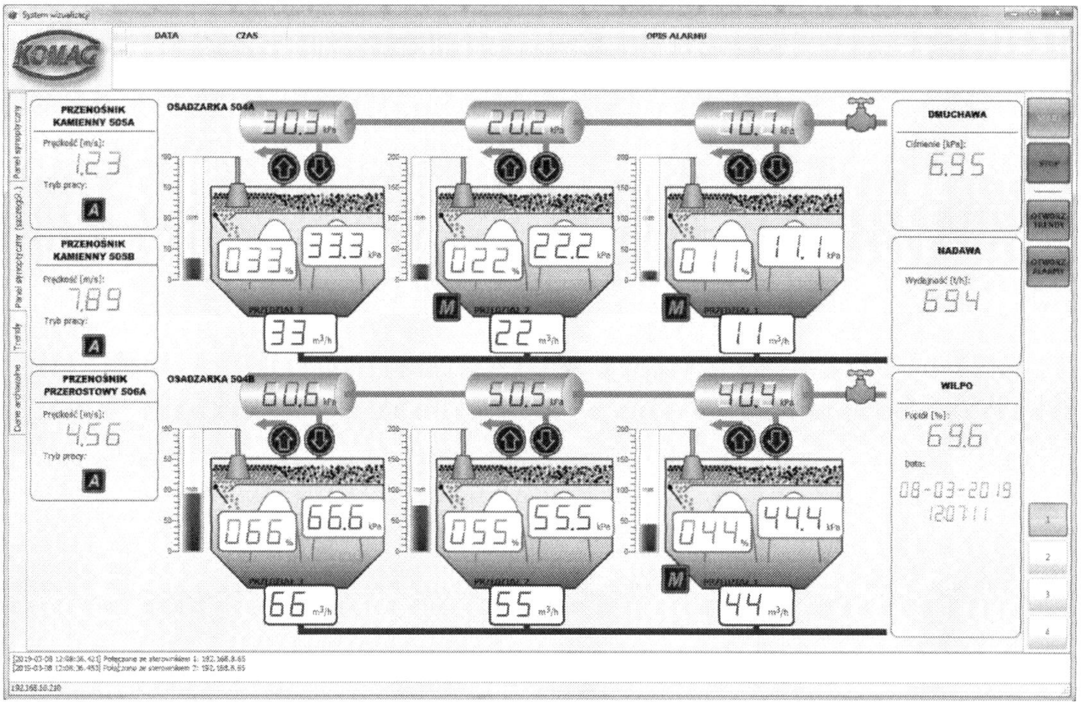

Figure 3. View of the visualization application window with operating parameters of jigs.

The application, beside the current visualization of the jig node operation, enables a presentation of time series of selected parameters and keeping the archives of measured parameters (figure 4). The window is divided into 6 parts, and each of them enables to display time series of parameters measured in the specific jig chambers. Changing of the displayed parameter or selecting the next one is performed by pressing the right mouse button on the selected interval and selecting a given range of parameters (figure 5).

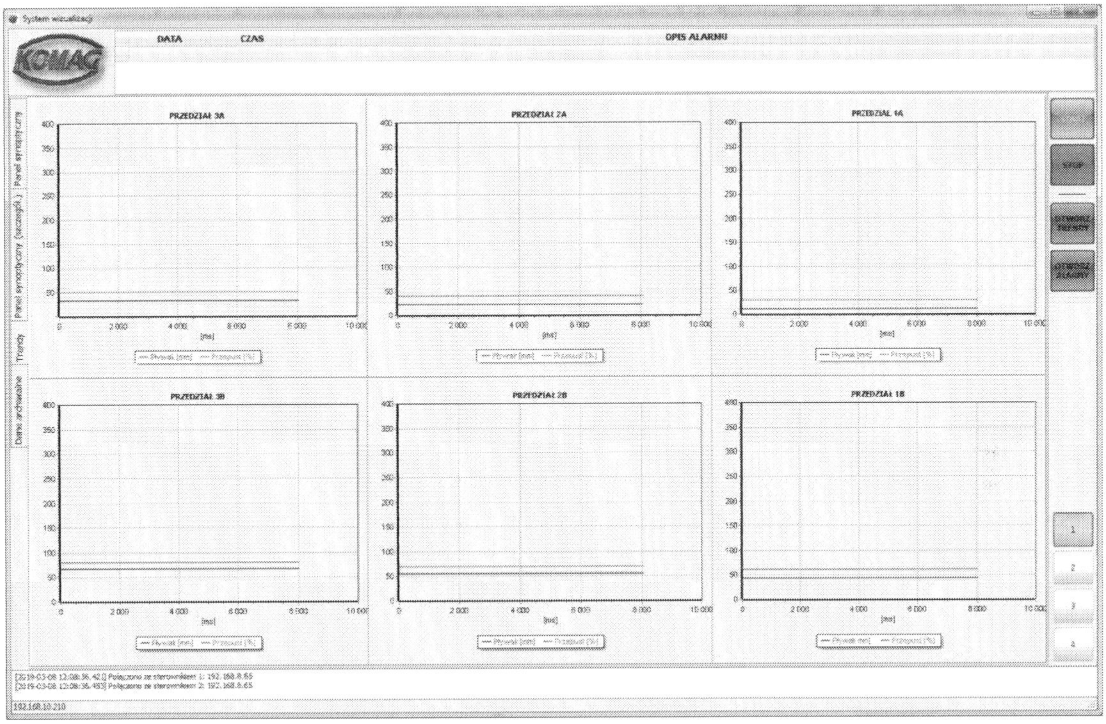

Figure 4. Visualization application window showing time series of selected parameters.

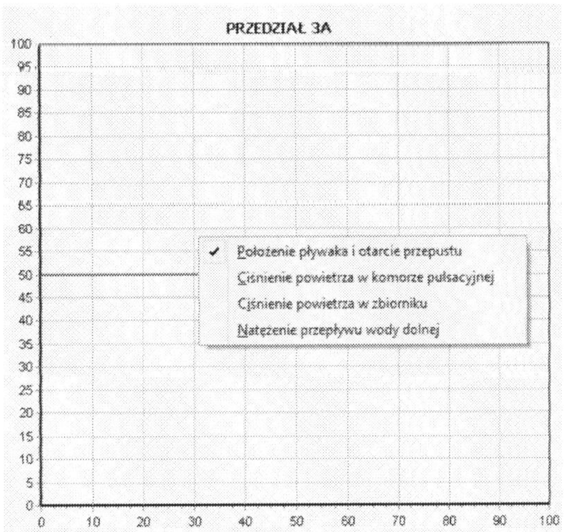

Figure 5. Trend window - range selection of displayed data.

An application also enables viewing alarms or time series of selected parameters from the last 30 days. The "Data" folder with historical data is created on an FTP server supported by the driver. With the start of a new day, further subdirectories are created in the "Data" directory, whose names correspond to the dates of their creation. Data files are saved daily in the subdirectories. A complete directory from the whole day contains 24 files with time series of selected parameters and one alarm file. Selecting a data file causes a display of the recorded waveforms in a tabular form (figure 6), as well as in a form of a graph (figure 7).

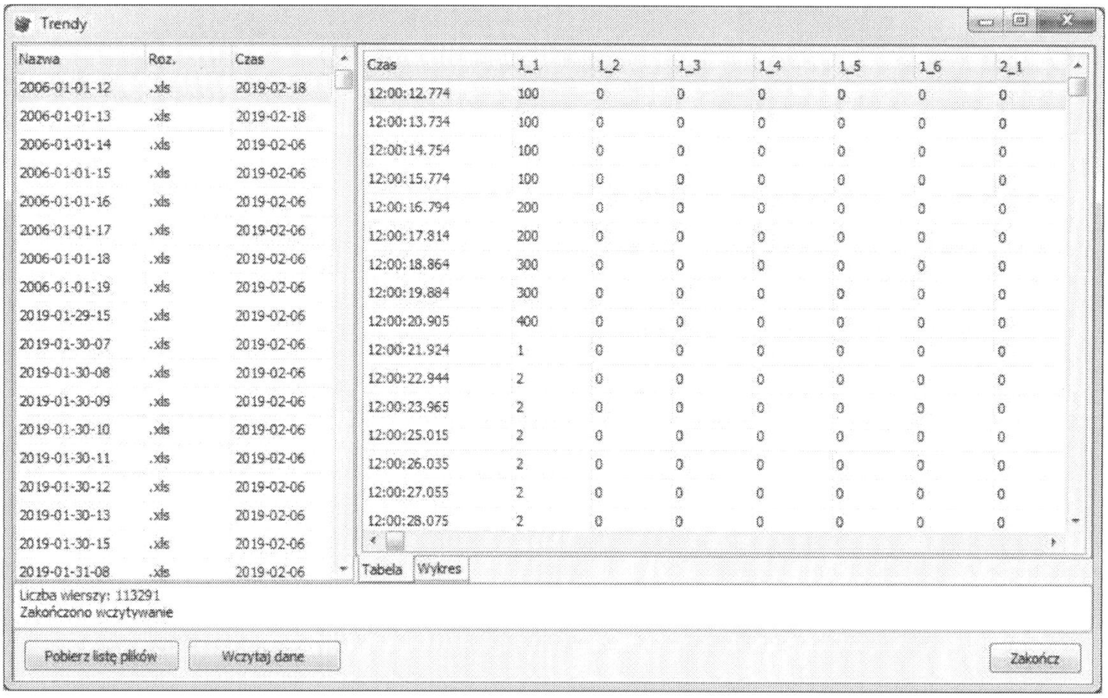

Figure 6. Visualization application window - archival data (in tabular form).

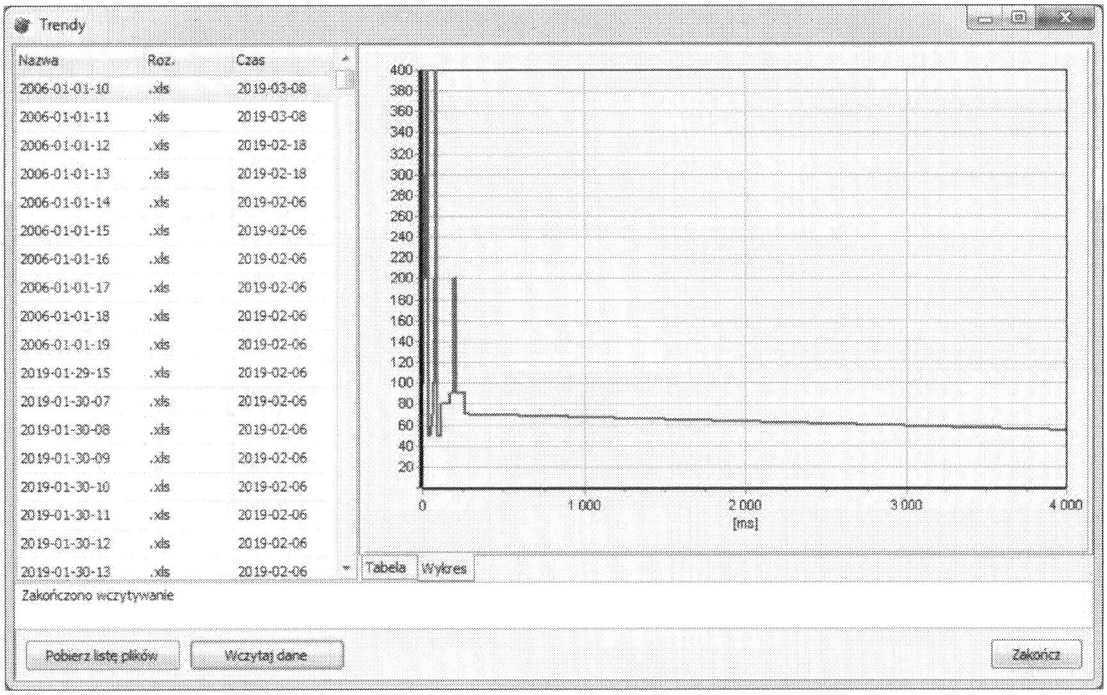

Figure 7. Visualization application window - archival data (in graphical form).

All the parameters are displayed on a single chart, which enables to compare the selected waveforms. It is possible to apply filters and select the data to be displayed on the chart. This is done by pressing the right mouse button and selecting the interesting parameters (figure 8).

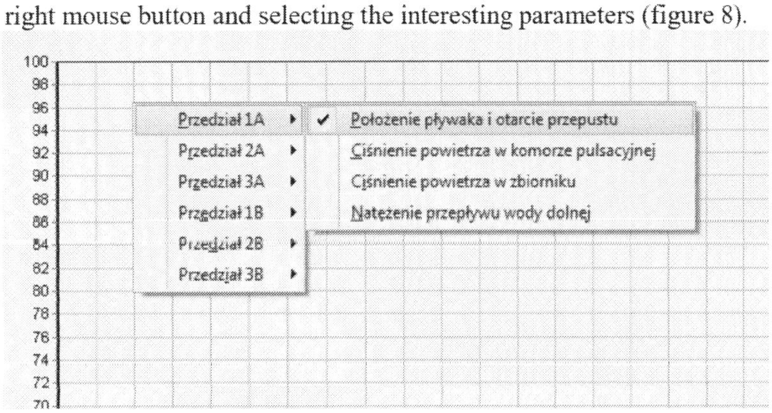

Figure 8. Archival trends window - filtering the range of displayed data.

A selection of the alarm file displays the alarms recorded on a given day (figure 9). It is possible to apply filters and display only the alarms from the selected ranges. This is realized by pressing the right mouse button and selecting an interesting ranges (figure 10).

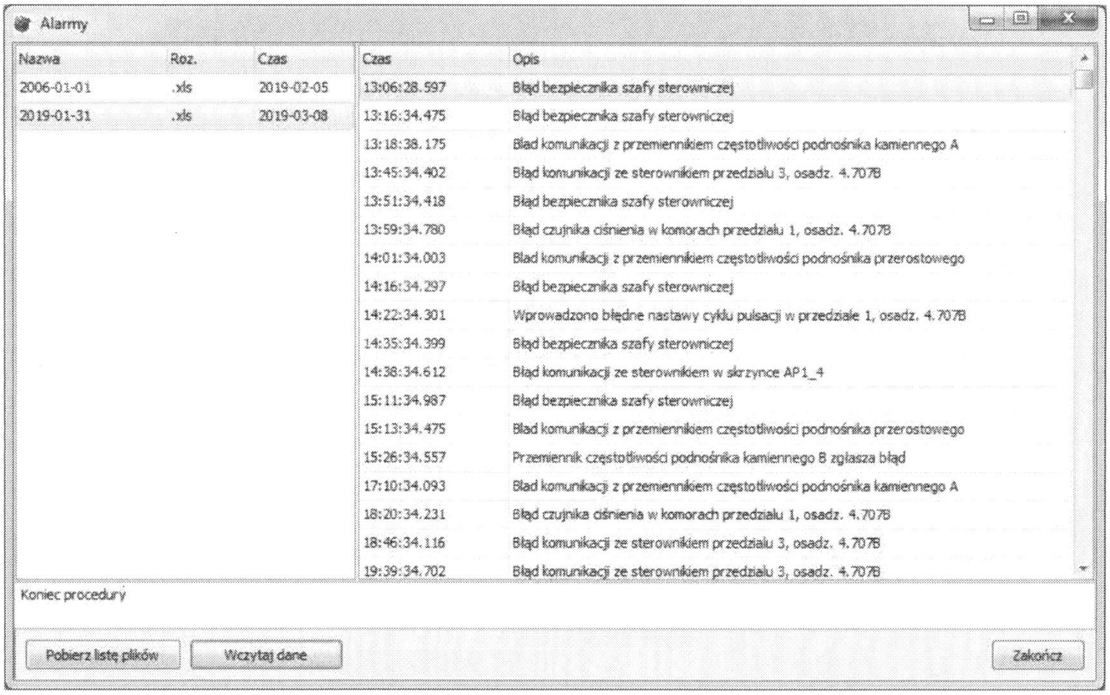

Figure 9. Window for reading recorded alarms.

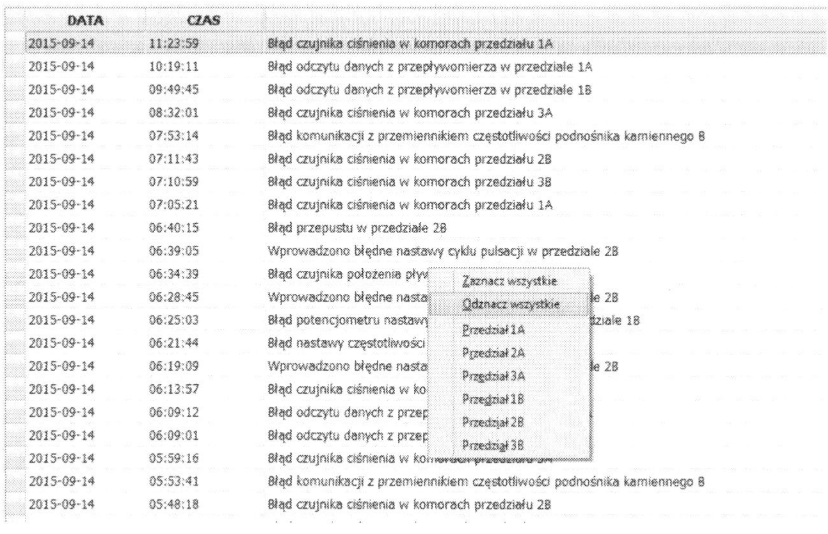

Figure 10. Archival alarm window - filtering the range of displayed alarms.

Since 2010, the correct and failure-free operation of the system installed in the processing plant of the mine, has been noticed.

The collected measurement data allow ITG KOMAG to carry out work related to search for further correlations of process parameters between individual machines included in the jig node [2, 3]. The algorithms implemented in the application enable advanced data analysis and contribute to the development of more perfect and reliable algorithms for effective control.

One of the tasks of the jig node control system is to stabilize the position of the separation layer (density of separation) by keeping the float at a given height. As part of the work carried out, an algorithm was implemented using a PID controller, which controls the opening of discharge slots through which bottom products are led out. In order to check the effectiveness of regulation, using author's application, the distributions of minimum position of the float in one of compartments of the jig were determined. Results were compared with the initial settings of the PID controller and after the tuning process using, among others, the author's application (figure 11 - 16).

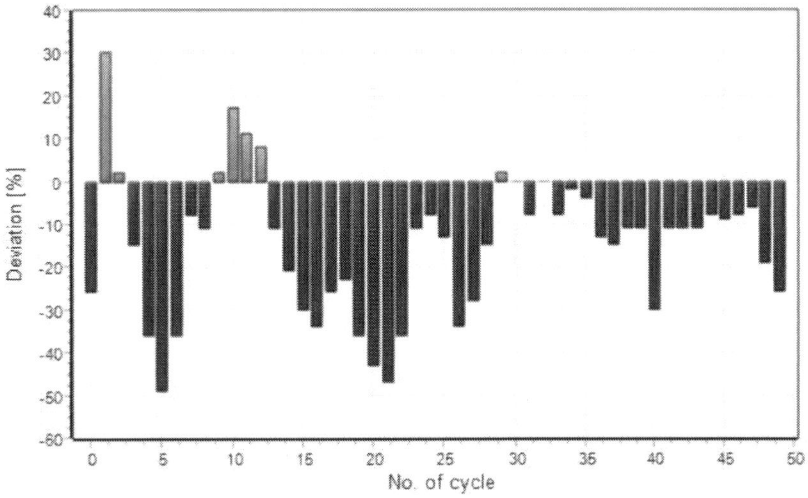

Figure 11. Fragment of the application window (diagram) - degree of deviation of the minimum position of float from the setpoint (PID regulator presetting).

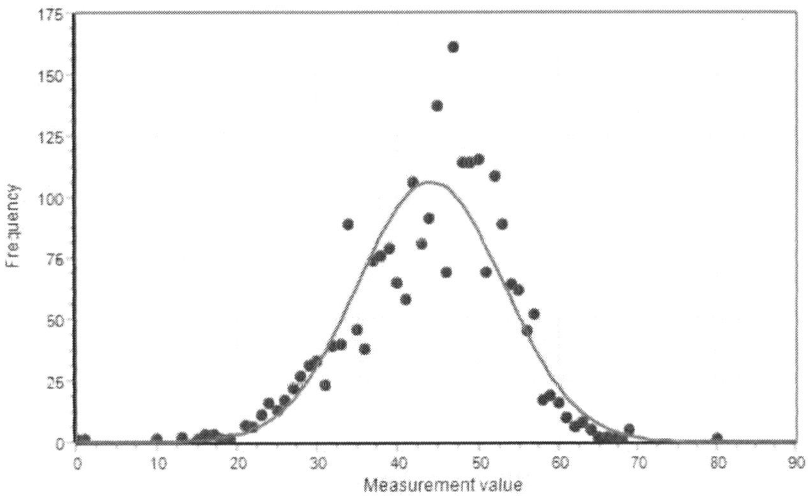

Figure 12. Fragment of application window (diagram) - distribution of change of minimum float position in the range no. 1 (PID regulator default settings).

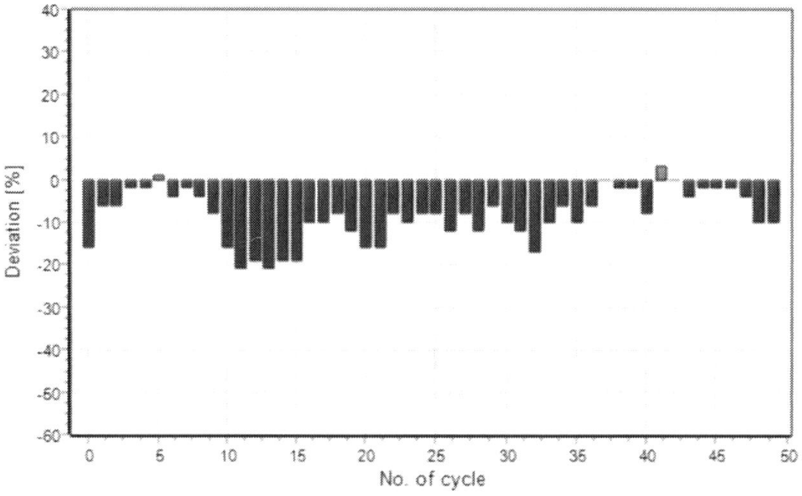

Figure 13. Fragment of the application window (diagram) - degree of deviation of the minimum position of the float from the setpoint (using the PID regulator - after 1st tuning).

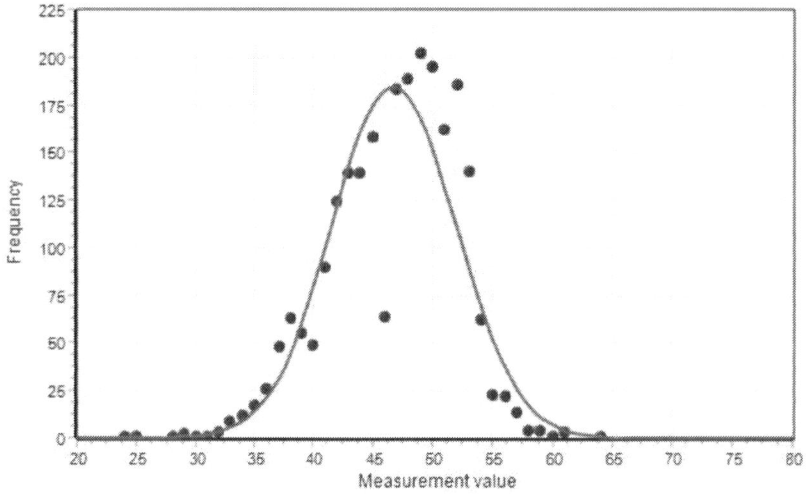

Figure 14. Fragment of application window (diagram) - distribution of minimum float position change in range no. 1 (using PID regulator - after 1st tuning).

IOP Conf. Series: Materials Science and Engineering **545** (2019) 012009 doi:10.1088/1757-899X/545/1/012009

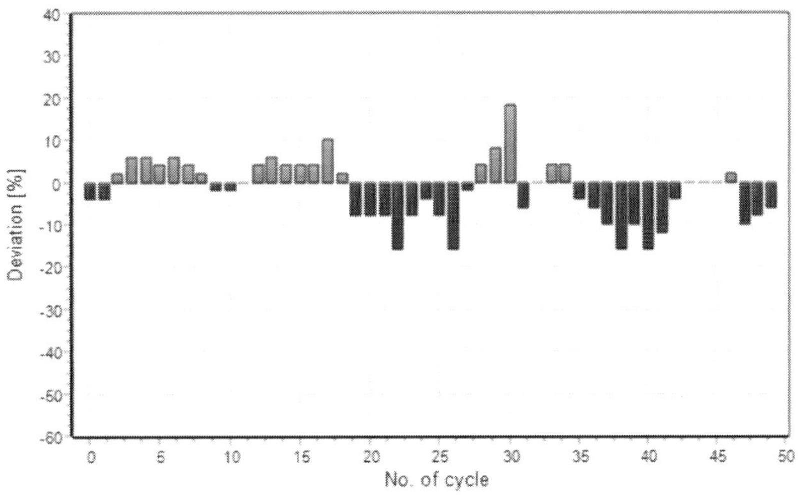

Figure 15. Fragment of the application window (diagram) - degree of deviation of the minimum position of float from the setpoint (with the use of the PID regulator - after 2nd tuning).

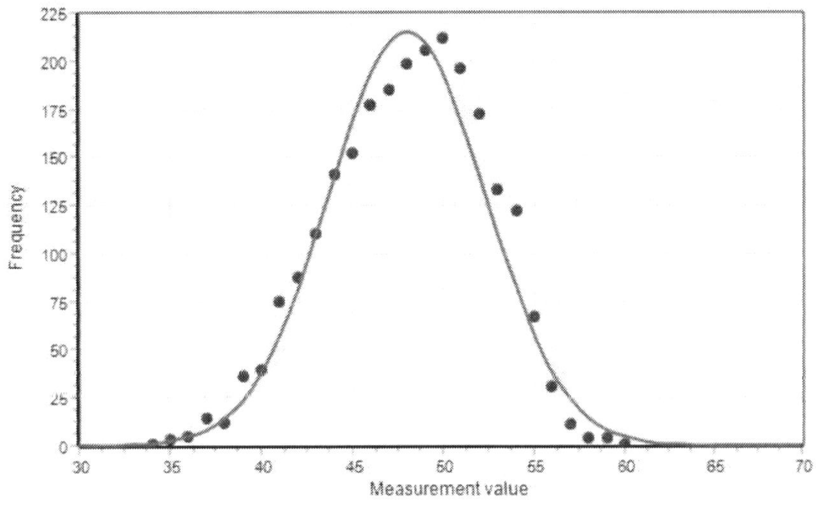

Figure 16. Fragment of application window (diagram) - distribution of change of minimum float position in range no. 1 (using PID regulator - after 2nd tuning).

The presented characteristics were recorded during three measurement sessions, in which the position of the float in one of the compartments was recorded for 60 minutes. The recorded mean values and standard deviations are presented in table 1.

Table 1. Minimum position of float without/with regulation system.

	PID controller presettings	With PID controller (after 1st tuning)	With PID controller (after 2nd tuning)
standard deviations	9,01	5,18	4,30
mean values	44,12	46,75	48,01

IOP Conf. Series: Materials Science and Engineering **545** (2019) 012009 doi:10.1088/1757-899X/545/1/012009

During the first batch in 60 minutes, the average value of the separation layer was 44.12 mm and the standard deviation was 9.01 mm. After the process of tuning PID controller, whose task was to maintain the separation layer at the set value of 53 mm, the mean value during the tests was 46.75 mm and the standard deviation was 5.18 mm. After another attempt to adjust the PID regulator settings, the mean value of 48.01 mm and the standard deviation of 4.30 mm were obtained. Comparing the distributions presented in figures 11-16, it can be concluded that the implementation of the control system stabilized the position of the distribution layer (distribution density) at the set value. The tests showed that the mean value was close to the setpoint. The regulator's operation was improved in terms of control efficiency.

As a part of further research and development work, it is planned to extend the application by an advisory module, which, basing on tracking short and long-term trends, will enable to predict hazards and optimize control algorithms. For this purpose, an interface will be implemented, which will enable creating, editing and deleting expert rules, used by the system, as well as procedures of concluding mechanism and of a visualization of its results.

4. Summary

The research and development work undertaken at KOMAG aims at an improvement of the existing solutions as well as a search for new opportunities to apply information technology. The article presents an example of a visualization system, operating together with a control system. The solution, presented above, has been implemented in mines and it is also used in the designing process as well as optimisation of control algorithms at the KOMAG Institute. As a part of further research and development work, it is planned to extend the application by an advisory module.

References

[1] Jendrysik S, Latos M, Jasiulek D, Stankiewicz K and Gawliński A 2015 Automatyczny system sterowania węzłem osadzarkowym z ciągłą kontrolą wydłużenia łańcucha przenośnika kubełkowego *Innowacyjne i przyjazne dla środowiska techniki i technologie przeróbki surowców mineralnych. Bezpieczeństwo - Jakość – Efektywność KOMEKO* pp 62-68

[2] Jendrysik S, Woszczyński M, Stankiewicz K and Gawliński A 2013 Układ sterowania węzłem osadzarkowym. KOMEKO 2013 Innowacyjne i przyjazne dla środowiska techniki i technologie przeróbki surowców mineralnych. Bezpieczeństwo - Jakość – Efektywność KOMEKO pp 179-187

[3] Cierpisz S and Kaula R 2014 Dobór parametrów regulatora dla obiektu inercyjnego z opóźnieniem na przykładzie osadzarki pulsacyjnej *Mechanizacja i Automatyzacja Górnictwa* **3**(517) pp5-13

[4] Jendrysik S, Woszczyński M and Jasiulek D 2017 Sterowanie procesem wzbogacania węgla w węzłach osadzarkowych *IV Polski Kongres Górniczy, Kraków* pp 531-537

[5] Król J, Krzak Ł, Jendrysik S, Stankiewicz K and Woszczyński M 2016 Wdrożenie układu sterowania węzłem osadzarkowym w KWK Sośnica *Innowacyjne i przyjazne dla środowiska techniki i technologie przeróbki surowców mineralnych. Bezpieczeństwo - Jakość – Efektywność KOMEKO* pp 227-237

[6] Kaula R 2015 Dobór nastaw regulatora PI w układach regulacji procesów wzbogacania węgla *Gospodarka Surowcami Mineralnymi – Mineral Resources Management* **31**(1) pp 141–154

IOP Conf. Series: Materials Science and Engineering **545** (2019) 012010 doi:10.1088/1757-899X/545/1/012010

Analysis of fraction distribution of the water drops stream generated by the spraying nozzles of new KOMAG design

D Bałaga M Kalita and M Siegmund

KOMAG Institute of Mining Technology, 37 Pszczyńska Street, 44-100 Gliwice, Poland

E-mail: msiegmund@komag.eu

Abstract. Theoretical backgrounds of airborne dust control with use of atomized stream of liquid are presented. Testing procedure, with a special attention paid to analysis of drops fraction distribution, is presented on the example of new nozzle design developed at KOMAG. The method for calculation of absorption surface is discussed. The test results are used to analyse operational parameters of the nozzle which are important for the dust control systems.

1. Introduction

A precipitation of dust particles with the use of atomized water stream is a commonly known method and it is often used both in the mining industry and other branches of industry. A combination of water drops generated by spraying nozzles with airborne dust particles is the main principle of dust control. There are several mechanisms of combining dust particles with water drops such as [1]:

- inertial,
- catching,
- diffusive.

When combining a water drop with a dust particle, its mass increases, what results in its faster settling on specially installed baffles (Figure 1) [2-4].

Figure 1. Settling of water drops with dust particles on specially installed baffles [2].

In order to increase the efficiency of combining dust particles with water drops, the area of dust absorption, i.e. the total surface area of all the drops of the spray stream should be maximized without changing their total volume.

An effectiveness assessment of the stream generated for a dust reduction systems should be carried out by analyzing the water drops fractional distribution in the spraying stream. The drops in the sprayed water stream make a typical heterogeneous (polydisperse) system, which is characterized by a large spread of drops diameters. An even larger spread occurs in drops surface areas and their masses. The polydisperse system, which is a sprayed water stream, can be presented graphically in a form of a volumetric distribution curve (volume share) in each range of drops diameters [1, 5-6]. To describe exactly a microstructure of the sprayed liquid, a total number of drops in a representative sample for analysis should be counted, measuring the diameters of the counted drops simultaneously. Then the total range of measured diameters should be divided into a dozen or so or several dozen diameter ranges, depending on the required accuracy and measurement method. Once the drop diameters, their number and volumetric in each interval are known, a quantitative, superficial or volumetric histogram is created. An example of a volumetric histogram is shown in Figure 2. By smoothing the stepped curve, showing the volumetric percentages of drops in individual diameter ranges, a volumetric distribution curve of drops with determined diameters can be obtained.

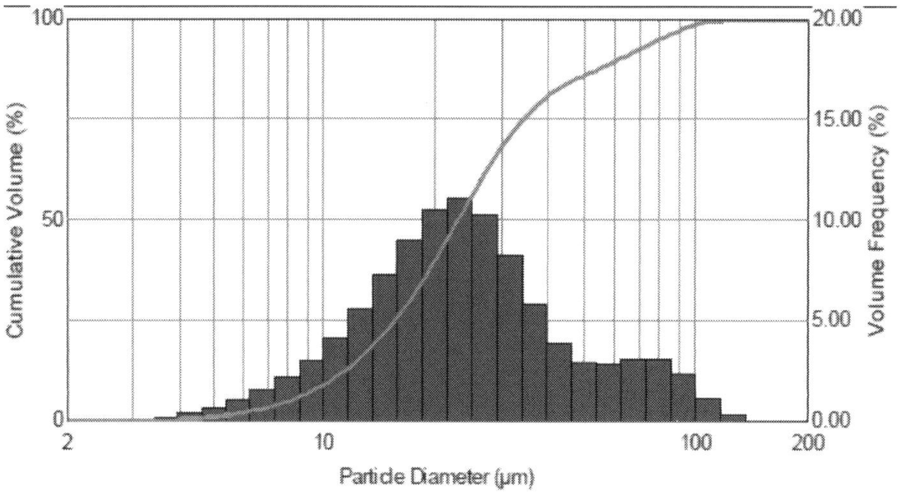

Figure 2. Exemplary of volumetric share of drops in each diameter range and the cumulative curve [1].

Due to a differentiation of drop sizes, it is necessary to characterize a taken drop sample by equivalent diameters. The equivalent diameter is an apparent quantity, which is characterized by a set of identical drops in place of the real drop population. The mean diameter, depending on the method of its calculation, determines the property of a set of drops such as: number, diameter, surface area and volume and it is a pictorial quantity used to assess a dispersion quality [5-7]. Due to a destination of the developed nozzle solution, i.e. dust control, in the part concerning an analysis of the test results, the D32 Sauter volumetric-superficial equivalent diameter was used. This diameter is the diameter of equivalent homogeneous set with the same total volume and the same total surface area of all the drops as in the real set (Figure 3) [5-7].

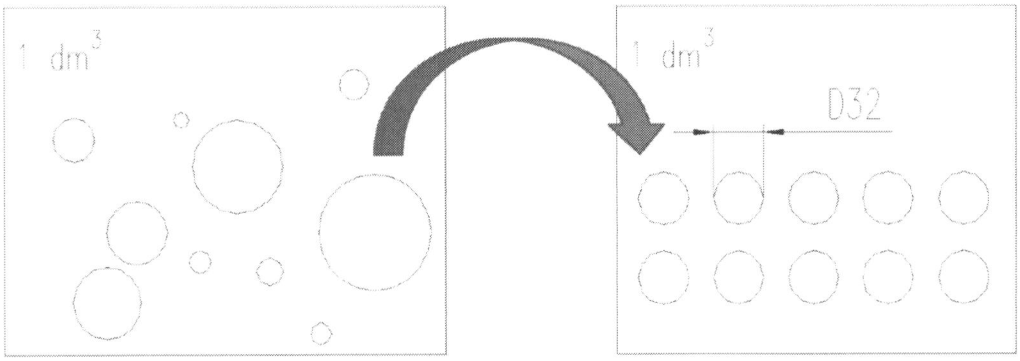

Figure 3. Illustration of Sauter's diameter definition [own report]

2. Absorption surface area calculation method

Knowing the Sauter's D32 equivalent diameter and the flow rate of media feeding the spray nozzle, we are able to calculate the total surface area of generated drops. In all the calculations, it is assumed that a drop has the shape of an ideal sphere whose surface area and volume is calculated from the following formula (2.1):

$$P_k = 4\pi \cdot r^3 \tag{2.1}$$

where:

P_k - surface area of a single drop, m^2,

r - drop radius (0.5 of D32 Sauter's diameter), m.

In turn, the volume of drops of D32 Sauter's diameter is calculated from the following formula (2.2):

$$V_k = \frac{4}{3}\pi \cdot r^3 \tag{2.2}$$

where:

V_k - volume of a single drop, m^3,

r - drop radius (0.5 of D32 Sauter's diameter), m.

A generated absorption surface area can be determined for a unit of time or spraying media flow rate. In the paper [8] authors suggested the following three variants of the parameter describing the spraying stream in form of absorption surface area of generated drops:

- **PA/T** - Absorption surface area of all the water drops in the spraying stream generated over time T equal to 1 min.
- **PA/W** - Absorption surface area of all the drops in the spraying stream generated at spraying of 1 dm^3 of water.
- **PA/P** - Absorption surface of all the drops in the spraying stream generated while using 1 Nm^3 of air.

Knowing the volumetric flow rate of water in the water pipeline feeding the nozzle Q_w and the volume of a single water drop according to the Sauter's equivalent diameter, we are able to calculate the equivalent number of all the drops in the spraying stream generated in a given time unit (min). The number of drops n_t is calculated from the following formula:

$$n_t = \frac{Q_w}{V_k} \tag{2.3}$$

where:
n_t - equivalent number of drops generated within 1 min,
Q_w - water flow rate, $\frac{\mathrm{m}^3}{\mathrm{min}}$,
V_k - volume of a single water drop, m^3.
The total absorption surface area generated by a nozzle during 1 minute is calculated from the formula:

$$PA/T = n_t \cdot P_k \tag{2.4}$$

where:

PA/T - absorption surface area generated within 1 min, m^2,

n_t - equivalent number of water drops generated *within 1 min,*

P_k - surface area of a single water *drop, m^2.*

Knowing the volume of a single water drop according to the Sauter's equivalent diameter, we are able to calculate the equivalent number of all the drops n_w in the spraying stream, generated from *1 dm^3* of water. The number of drops n_w is calculated from the following formula:

$$n_w = \frac{1\ dm^3}{V_k} \tag{2.5}$$

where:
n_w - equivalent number of water drops generated from *1 dm^3* of water,
V_k - volume of a single drop, m^3.
The total absorption surface area generated by a nozzle from 1 dm^3 of water is calculated from the following formula:

$$PA/W = n_w \cdot P_k \tag{2.6}$$

where:
PA/W - absorption surface area generated from 1 dm^3 of water, m^2,
n_w - equivalent number of drops generated from 1 dm^3 of water,
P_k - surface area of a single water drop, m^2.
To calculate the equivalent surface area of all the water drops generated from 1 Nm^3 of air, the time t of nozzle operation, required for using 1 Nm^3 of air, should be specified. It is calculated from the following formula:

$$t = \frac{1\ Nm^3}{Q_p} \tag{2.7}$$

where:
t - time of nozzle operation required for *using 1 Nm^3* of air, min,
Q_w - air flow rate, $\frac{\mathrm{Nm}^3}{\mathrm{min}}$,
The number of equivalent water drops n_p , generated by the nozzle during the time needed for using 1 Nm^3 of air, is calculated from the formula:

$$n_{wp} = t \cdot n_t \tag{2.8}$$

IOP Conf. Series: Materials Science and Engineering **545** (2019) 012010 doi:10.1088/1757-899X/545/1/012010

where:

n_p - equivalent number of water drops generated from 1 Nm^3 of air,

t - time of nozzle operation required for using 1 Nm^3 of air, min,

n_t - equivalent number of drops generated within 1 min,

The total absorption surface area of water drops generated by a nozzle from 1 Nm^3 of air, is calculated from the following formula:

$$PA/P = n_p \cdot P_k \tag{2.9}$$

where:

PA/P - absorption surface area generated from 1 Nm^3 of air, m^2,

n_p - equivalent number of water drops generated from 1 Nm^3 of air,

P_k - surface area of a single water drop, m^2.

3. Test of operational parameters of the spraying nozzle designed at KOMAG

Testing the operational parameters of a new design spraying nozzle consisted in a drops fraction distribution analysis as well as in a consumption of spraying media in relation to assumed supply pressure. A self-cleaning DSD type nozzle, presented in Figure 4, was tested.

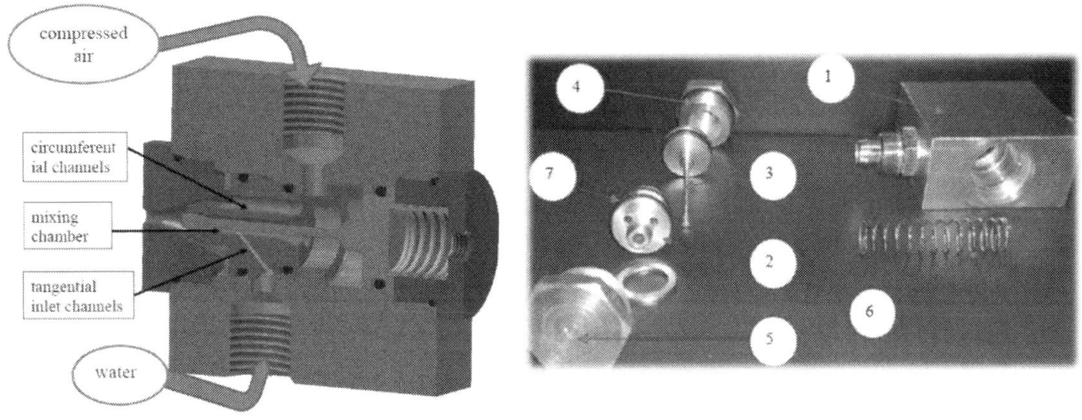

Figure 4. Self-cleaning nozzle components: 1 - connecting body, 2 - internal nozzle, 3 - needle with piston, 4 - outlet sleeve, 5 - plug, 6 - spring, 7 - seals [9].

Tests of operational parameters of the DSD self-cleaning nozzle were conducted according to the accepted testing methodology. They consisted in recording, using the appropriate equipment, the following parameters:

- distribution of drops diameters in the sprayed liquid stream,
- supply pressure and the volumetric air flow rate in the air pipeline, supplying the nozzle,
- supply pressure and the volumetric water flow rate in the water pipeline, supplying the nozzle,

The schematic diagram of the test stand is shown in Figure 5, and Figure 6 shows the view of this stand during the measurements.

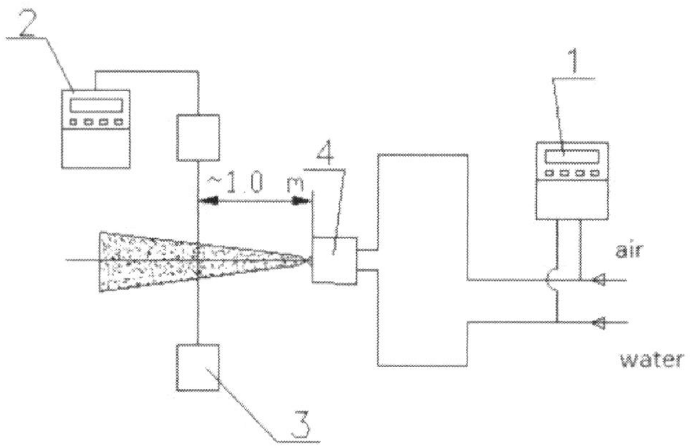

Fig. 5. Test stand (Bałaga, Jaszczuk, 2016; Siegmund, 2017): 1- media measuring system (water and air), 2- recording equipment of drops analyser, 3- drops analyser, 4- air-water nozzle with fixing body.

Figure 6. View of the test stand during the measurement of fractional distribution of drops generated by DSD nozzle [9].

In the DSD nozzle tests, measurements were taken for the same water and air pressures, i.e.; 0.3; 0.4; 0.5 and 0.6 MPa. The same pressures of water supply and compressed air resulted from by the nozzle design, which ensured its correct operation only in the case of the same pressure values. The nozzle construction enables its adaptation to a few variants of water supply (by using a different number of water inlet holes). It is also possible to operate a nozzle without a cleaning needle.

The DSD nozzle tests were carried out for the following configurations of mixing chamber supply and outlet cleaning system:

three 0.8 mm dia water inlet holes with a closing needle (nozzle type; DSD - K1),

one 0.8 mm dia water inlet hole, without a closing needle (nozzle type; DSD - K2),

one 0.8 mm dia water inlet hole, with a closing needle (nozzle type; DSD -K3).

The results of testing the spray stream parameters for each K configuration are presented in Table 1. Basing on the measured water and compressed air flowrate as well as D32 Sauter's equivalent diameter, the absorption surface area of the generated spraying stream was calculated.

Table 1. Test results of various self-cleaning nozzle configurations [9].

DSD – K1 - configuration 1- 3x 0.8 mm inlet hole,
DSD – K2 - configuration 2- 1x 0.8 mm inlet hole without the cleaning needle,
DSD –K3 - configuration 3- 1x 0.8 mm inlet hole with the cleaning needle.

Nozzle type	Pressure of supplying media water/air [MPa/MPa]	Water flow rate [dm^3/min]	Air flow rate [Nm^3/min]	D32 Sauter's diameter [μm]	Absorption surface area PA/T [m^2]	Absorption surface area PA/W [m^2]	Absorption surface area PA/P [m^2]
DSD-K1	**0.3/0.3**	0.32	0.042	43	44.7	139.5	1063.1
	0.4/0.4	0.41	0.062	41	60.0	146.3	967.7
	0.5/0.5	0.5	0.072	41	73.2	146.3	1016.3
	0.6/0.6	0.64	0.081	39.8	96.5	150.8	1191.1
DSD-K2	**0.3/0.3**	0.2	0.072	21	57.1	285.7	793.7
	0.4/0.4	0.22	0.103	19	69.5	315.8	674.5
	0.5/0.5	0.24	0.129	18	80.0	333.3	620.2
	0.6/0.6	0.3	0.181	17	105.9	352.9	585.0
DSD-K3	**0.3/0.3**	0.15	0.059	21	42.9	285.7	726.4
	0.4/0.4	0.2	0.089	18	66.7	333.3	749.1
	0.5/0.5	0.22	0.116	16	82.5	375.0	711.2
	0.6/0.6	0.24	0.144	16	90.0	375.0	625.0

Basing on the data from Table 1, comparative graphs for all the tested DSD self-cleaning nozzle configurations were developed. Figure 7 shows a graph of changes in the water flow rate in relation to the pressure of the supplying media. The volumetric water flow in the nozzle with three water inlet holes configuration was about 2 times larger than the flow for the nozzle with one water inlet hole configuration, irrespective of the cleaning needle installation. It should be noted that its absence has a minor effect on the water flow rate increase in relation to configuration with the cleaning needle.

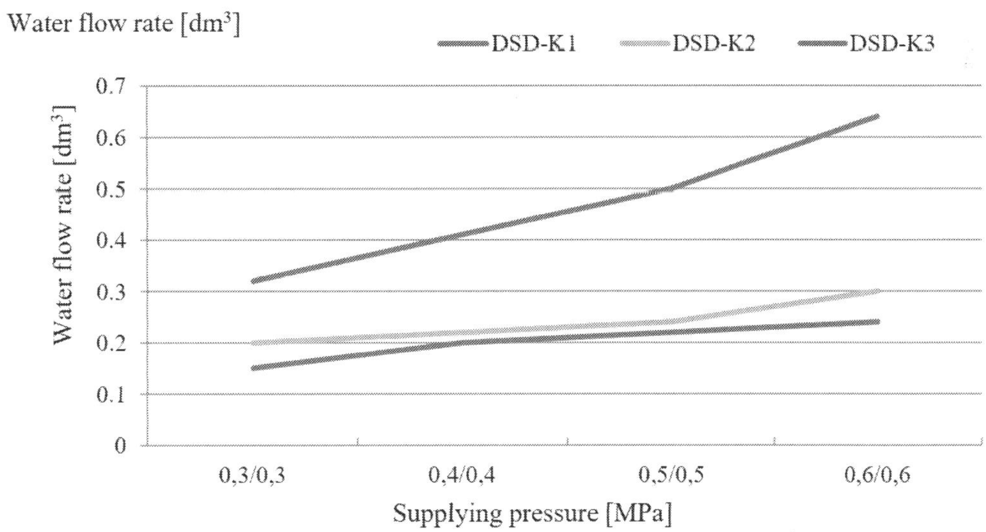

Figure 7. Measurement of fractional distribution of drops diameters [9].

IOP Conf. Series: Materials Science and Engineering **545** (2019) 012010 doi:10.1088/1757-899X/545/1/012010

Figure 8 shows a graph of changes in the basic quality parameter of the spraying stream, which is the D32 Sauter's equivalent diameter, in relation to the supplying media pressures. The D32 diameter at the three water inlet holes, for different pressure values, varied between 40 and 45 μm. Supplying water through only 1 inlet hole resulted in a generation of a spraying stream in the range of water unit <20 μm. An impact of the cleaning needle on a fractional distribution of drops in the stream is practically unnoticeable.

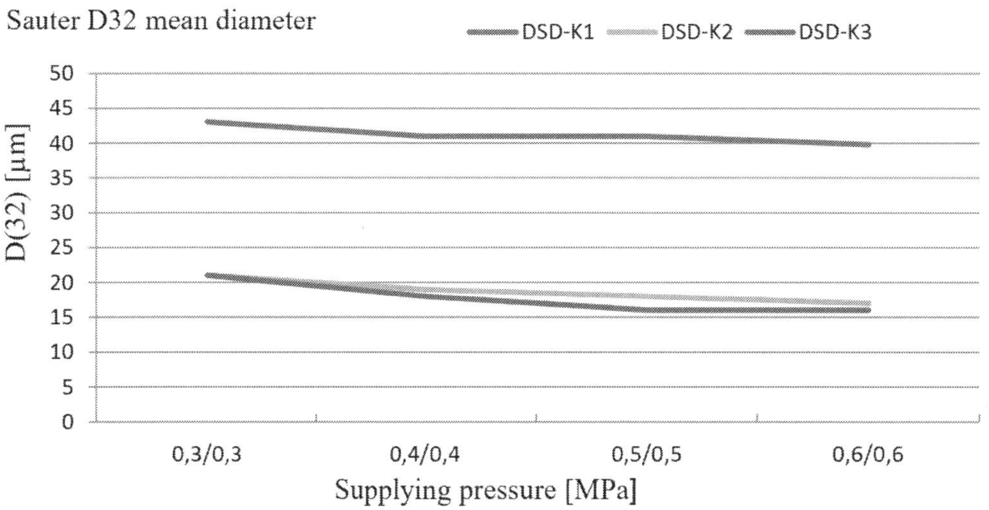

Fig. 8. Characteristic (D32) equivalent diameter in relation to supplying media pressures [9].

Fig. 9 shows a graph of changes in the PA/T absorption surface areas in relation to the supplying media pressures. All the configurations of the self-cleaning nozzle tend to increase the absorption surface areas proportionally to supplying pressure increase. The configuration of the nozzle with three water inlet holes, despite of using about twice the volume of water, generated the smallest absorption surface area. An impact of the cleaning needle is only important for extreme supplying pressures, where the area of absorbed surfaces is reduced in relation to the configuration without the cleaning needle.

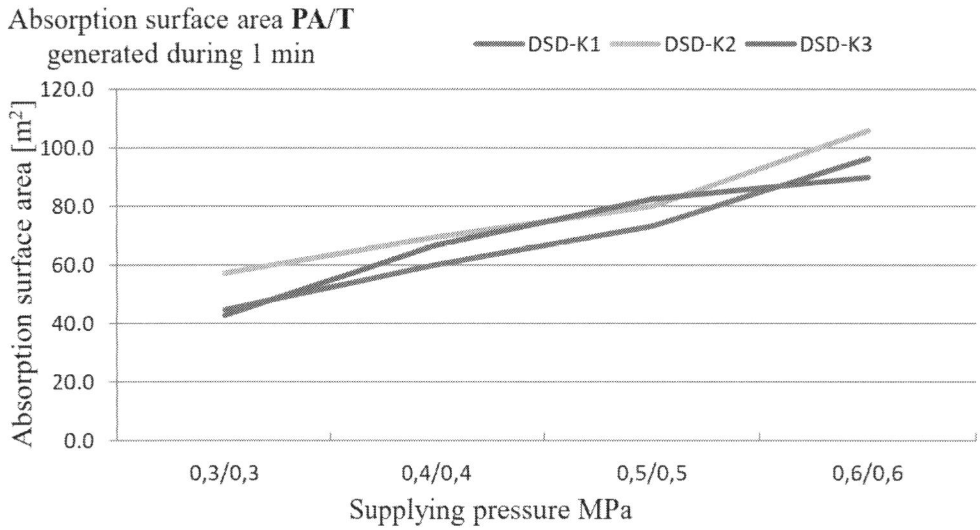

Figure 9. Absorption surface area PA/T in relation to supplying media pressure [9].

Figure 10 shows a graph of changes in PA/W absorption surface areas in relation to the supplying media pressures. Comparing the areas of the absorption surface generated from one dm^3 of water, a much higher efficiency of the nozzles in the configuration with one water inlet hole can be noticed. They produce the absorption surface area about 2.5 times larger.

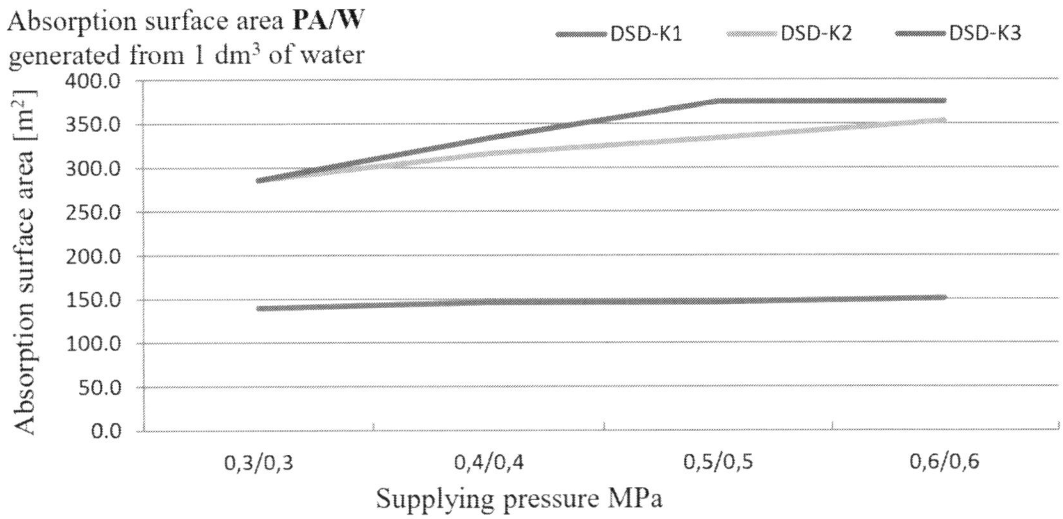

Figure 10. Absorption surface area PA/W in relation to supplying media pressure [9].

Figure 11 shows a graph of changes in PA/P absorption areas in relation to the supplying media pressures. Comparing the absorption surface areas, generated from 1 Nm^3, we observed 60-70 % higher efficiency of the nozzle in K1 configuration in relation to K2 and K3 configurations

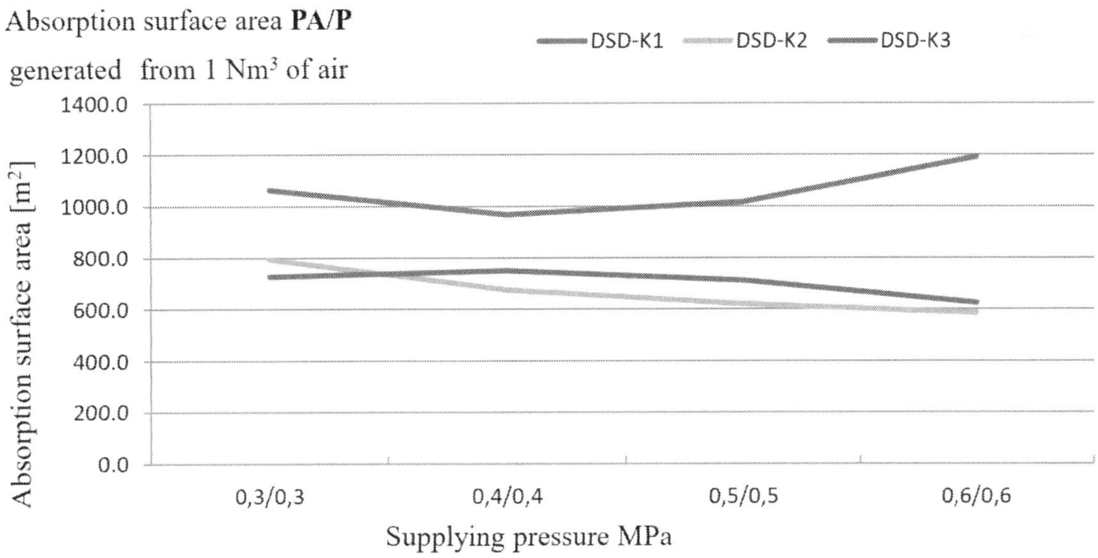

Figure 11. Absorption surface area PA/P in relation to supplying media pressures [9]

IOP Conf. Series: Materials Science and Engineering **545** (2019) 012010 doi:10.1088/1757-899X/545/1/012010

4. Summary and conclusions

The presented nozzle assessment procedure in terms of a degree of water drops atomisation allows for a determination of equivalent absorption surface areas of the spraying water stream. An introduction of parameters, directly specifying the total surface area of the generated drops to a common engineering practice, simplifies a process of selecting nozzles for spraying systems [10]. The suggested parameters, describing an absorption surface area of drops in the generated stream should, according to the authors, be included in nozzle datasheets, as they provide information about nozzle operational parameters and indirectly about their dust control effectiveness. This parameter, in the author's opinion, will facilitate a selection of a correct type of a spraying nozzle for a designed spraying installation made by designers and users of spraying systems.

During the tests of the DSD type spraying nozzles it was found as follows:

- a high degree of atomisation of the spraying stream in the result of using only one 0.8 mm dia water inlet hole,
- insignificant impact of the needle on the distribution of the D32 characteristic diameter,
- a nozzle with three 0.8 mm dia water inlet holes, generated in the same time, the same absorption surface area as a nozzle with one 0.8 mm dia water inlet hole, at the water consumption 2 ÷ 2.5 times smaller.

References

[1] Bałaga D and Jaszczuk M 2016 *Wpływ parametrów strumienia zraszającego na redukcję zapylenia generowanego przez kombajn ścianowy* Prace Naukowe - Monografie *KOMAG* **no 47**, pp.1-107

[2] Bałaga D et al. 2015 Powietrzno-wodne instalacje zraszające do redukcji zapylenia *Wiad. Gór.* **11** p 560-568.

[3] Prostański D 2012 Dust Control with Use of Air-Water Spraying System *Arch. Min. Sci.* **4** pp 975-90

[4] Prostański D 2018 Development of research work in the air-water spraying area for reduction of methane and coal dust explosion hazard as well as for dust control in the Polish mining industry *IOP Conf. Ser.: Mater. Sci. Eng.* **427** 012026

[5] Orzechowski Z and Prywer J 1991 *Rozpylanie cieczy* Wydawnictwo Naukowo Techniczne, Warszawa

[6] Orzechowski Z and Prywer J 2008 *Wytwarzanie i zastosowanie rozpylonej cieczy* Wydawnictwo Naukowo Techniczne Warszawa.

[7] Zbrożek P and Prasuła J 2009 Wpływ wielkości średnic kropli mgły wodnej na efektywność tłumienia pożarów i chłodzenie *Bezpieczeństwo i Technika pożarnicza* **3** p 113-148.

[8] Bałaga D Kalita M and Siegmund M 2018 Badania parametrów strumieni zraszających dysz drobnokroplistych *Masz. Gór.* **3** p 3-13

[9] Siegmund M 2017 *Drobnokroplista, samoczyszcząca dysza powietrzno-wodna* Report M-EM17-22485/OR (unpublished materials), KOMAG Gliwice

[10] Prostański D 2017 *Zraszanie powietrzno-wodne jako metoda ograniczenia zagrożenia zapłonem metanu i wybuchem pyłu węglowego oraz redukcji zapylenia powietrza* Prace Naukowe - Monografie KOMAG no 51 pp. 1-193

IOP Conf. Series: Materials Science and Engineering **545** (2019) 012011 doi:10.1088/1757-899X/545/1/012011

Bench tests of BWC pressure free heat exchanger - part 1

H Suffner and K Lesiak

[1]KOMAG Institute of Mining Technology, Division of Machines and Equipment, 37 Pszczyńska Street, 44-101 Gliwice, Poland

hsuffner@komag.eu, klesiak@komag.eu

Abstract: A requirement to limit the temperature of diesel engine components results from the recommendations of the drive manufacturers as well as from the mining regulations. In order to meet these requirements it is necessary to reduce the temperature of the exhaust gases below 150 °C and in order to achieve this objective, in the exhaust system of the diesel engine various types of heat exchangers are used. The article presents the procedures and results of testing the pressure free heat exchanger for cooling the exhaust systems in underground mining machinery drives applied in hard coal mines.

1. Introduction

A limitation of exhaust gases temperature is one of the requirements for diesel engines operated in hard coal as well as metal ores and other minerals mines. During the operation of a heat machine, which is a piston diesel engine, the thermal energy is converted into mechanical energy. Unfortunately, only a part of the fed thermal energy is converted into work, while a significant part is emitted to the environment [3].

Problems related to cooling of exhaust gases in diesel engines used in hard coal mines and a development of the devices designed for this purpose are discussed in the article [5].

The article presents the assumptions, testing procedure and results of testing the pressure-free heat exchanger (PFHE) on the test stand at KOMAG in Gliwice.

2. Reason for undertaking a realization of a pressure free heat exchanger design

Two solutions, i.e. wet scrubbers and dry heat exchangers, are used to cool exhaust gases from diesel engines operating in underground workings of hard coal mines. In the wet scrubbers, the stream of diesel exhaust gases is directed to the box filled with water. The gases, flowing through water, transfer a part of heat to water and the resulting mixture of exhaust gases, steam and water drops are further directed to the next part of the exhaust system and then to the atmosphere. In the dry heat exchanger the exhaust gases are directed to a chamber made of the pipe system in which cooling agent circulates. Heat of exhaust gases is transferred to the cooling agent through the pipes wall. The wet scrubber, despite its better efficiency in cooling the exhaust gases, has a big disadvantage of a continuous loss of water in the cooling chamber.

The water is partly transferred into a steam and flows out together with gases and partly it flows out as small water drops entrained by exhaust gases. That is why water has to be periodically refilled in the cooling box. Depending on an intensity of using the diesel engine, water should be refilled once or twice per shift.

Users of diesel engines with wet scrubbers draw attention to a necessity of limiting the above mentioned water refilling process. This process disorganizes an operation of the machine with a diesel drive and water torn out from the exhaust system is wasted as it cools down the gases only partially.

KOMAG specialists have undertaken many research activities aiming at a limitation of volume of water torn from the wet scrubber by e.g. changing the scrubber box design and introducing a spiral outlet with bouncing metal sheets [1].

Recently, a project on a pressure free heat exchanger was realized, an experimental version of such a device was developed [6] and the stand tests of it were conducted [4].

A pressure free heat exchanger is a device combining the advantages of a wet scrubber and of a dry heat exchanger. In this device, the exhaust gases flow in the system of pipes immersed in water, transferring a part of heat to water causing its evaporation. The outlet of exhaust gases from the pipes system is placed over the water surface, where the gases combine with steam and flow to the outlet duct. The gases are cooled down during the flow through the outlet duct. An efficiency of cooling the exhaust gases is nearly the same as the cooling efficiency of a wet scrubber but in this case the water drops are not torn out from the system .

In Figure 1 a sketch of a pressure free heat exchanger (PFHE) is presented.

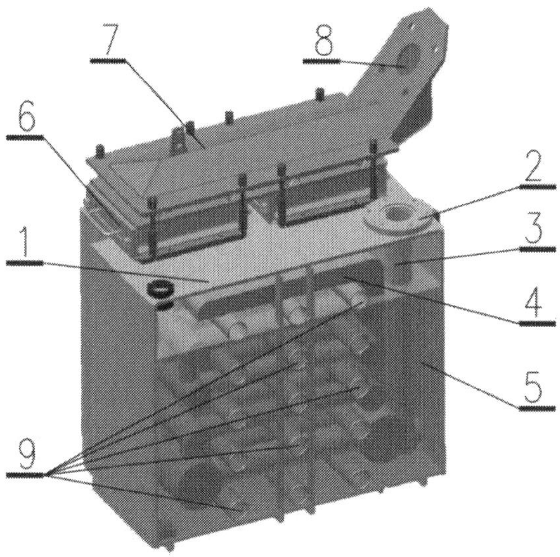

Figure 1. Pressure free heat exchanger: 1 – heat exchanger box, 2 – connection of the exhaust gases outlet, 3 – exhaust gases outlet duct, 4 – labyrinth of cooling pipes, 5 – water, 6 – flame arrester, 7 – exhaust manifold, 8 – connection of spark arrester, 9 – strengthening pipes [own source].

3. Testing assumptions (objectives)

A verification of the pressure free heat exchanger efficiency i.e. an ability of cooling the exhaust gases and transferring the heat to the atmosphere was the main tests objective. A comparison of the PFHE efficiency and of the wet scrubber according to KOMAG design was the second tests objective. During the designing process it was assumed that a new design was to be compatible with the design of wet scrubbers used in underground locomotives designed at KOMAG to be replaced by PFHE during an overhaul if the users wanted to do this.

It was assumed that an ability of cooling the exhaust gases and transferring the heat to the atmosphere would be assessed basing on temperature measurements at selected points of diesel drive exhaust system. The assumption is in line with one of the requirements put upon diesel drives, specifying the maximum unit surface temperature of the diesel drive.

4. Testing stand and testing object

The tests were started on the test stand designed for testing KOMAG diesel drives with a heat exchanger in a form of a wet scrubber to realize the second testing objective (a comparison of PFHE efficiency with the exhaust gases wet scrubber according to KOMAG solution).

It was assumed that the tests of PFHE would be conducted on a modernized stand for testing diesel drives.

The modernization consisted in an adaptation of the stand for an installation of the PFHE and equipping it with additional temperature sensors used for an assessment of the PFHE efficiency.

In Figure 2 a model of the test stand for diesel drives with the installed PFHE is shown.

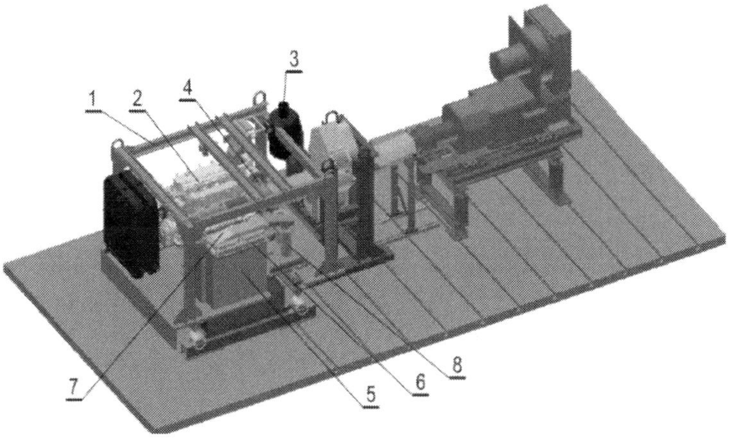

Figure 2. Model of the test stand for diesel drives: 1 – fixing frame; 2 – diesel engine; 3 – inlet system; 4 – exhaust gases outlet duct; 5 – pressure free heat exchanger (PFHE); 6 – flame arrester; 7 – exhaust manifold; 8 – spark arrester or exhaust gases outlet duct [own source].

D5AT diesel engine of 81 kW power was installed on the test stand. The engine characteristics is given in Fig.3.

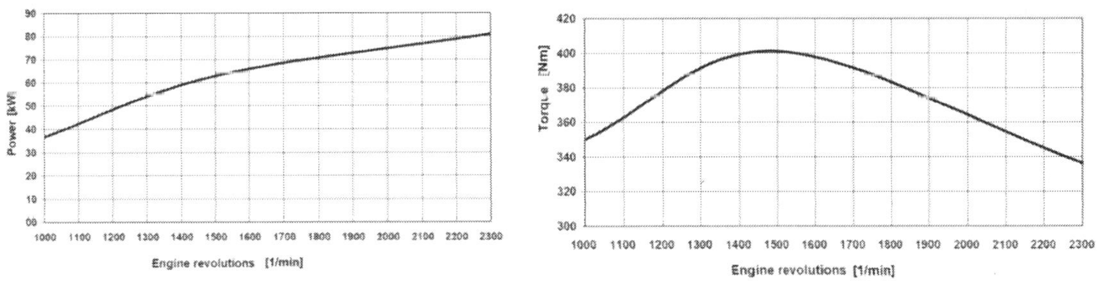

Figure 3. Characteristics of D5AT engine operation.

An experimental version of the pressure free heat exchanger PFHE was the testing object. It differs from the commercial version as it is not in the explosion-proof design and it is manufactured as an easy-to-be-disassembled box without any additional enhancements. Such a solution resulted from an intension to test two versions of cooling pipes labyrinth (see Fig. 1), hereinafter called heat exchangers. These two versions differ by the parameters of used cooling pipes. In Figure 4 the heat exchanger with pipes d=44.5 mm is shown.

IOP Conf. Series: Materials Science and Engineering **545** (2019) 012011 doi:10.1088/1757-899X/545/1/012011

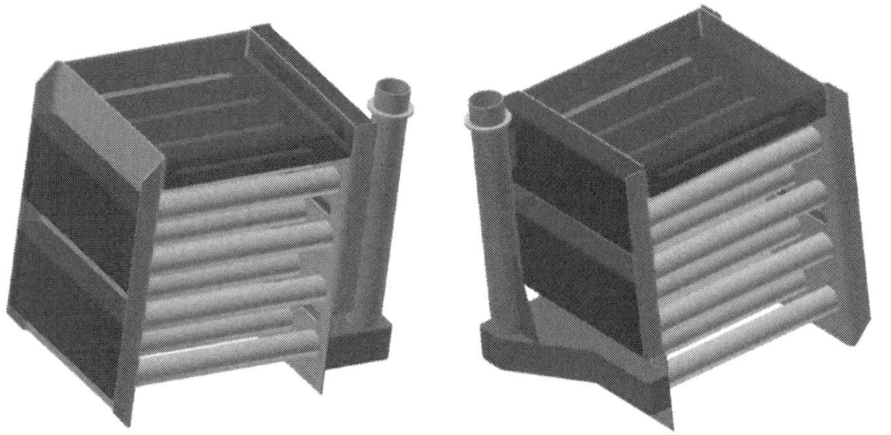

Figure 4. Heat exchanger d=44.5 mm.

For the second heat exchanger version with pipes d=20 mm, the following two testing cycles were planned: the standard one as for the heat exchanger 44.5 mm and the cycle modelling the partial clogging of the cooling pipes. The effect of partial clogging was obtained by blocking a part of the pipes using special plugs. The heat exchanger d=20 is shown in Fig. 5.

Heat exchanger with dismounted cover plate and mounted plugs

Figure 5. Heat exchanger d=20 mm.

Testing procedure

The PFHE efficiency tests consisted in measuring and recording temperatures at specific points of the exhaust gas outlet system, as well as in measuring the water level in the heat exchanger.
A list of measurement points is given in Table 1.

Table 1. List of temperature measurement points and cooling water surface level.

Symbol	Description	Fixation
P1	In the stream of gases flowing out of the turbocharger	socket M12x1,5
P2	Surface of the outlet duct behind the turbocharger	socket M12x1,5
P3	In the stream of gases flowing into the heat exchanger	socket M8
P4	Surface of the outlet duct before the heat exchanger	glued sensor
P5 and 5a	The external surface of the heat exchanger box	glued sensor
P6a and P6b	Water temperature inside the heat exchanger box	2 sockets M12x1,5
P7a P7b	The temperature of the mixture of exhaust gas and steam at the outlet of the heat exchanger	2 sockets M12x1,5
P8a and P8b	The front surface of flame arresters	glued sensor
P9	Surface of the spark arrestor (if present)	glued sensor
P10	Surface of the exhaust manifold (behind arresters)	glued sensor
P11	Surface of the outlet duct	glued sensor
P12	Exhaust gases outlet	socket M12x1,5

Figure 6 shows an arrangement of measuring points on the tested object.

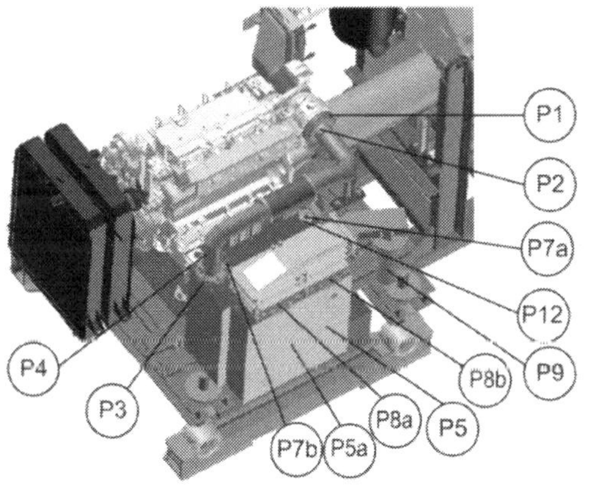

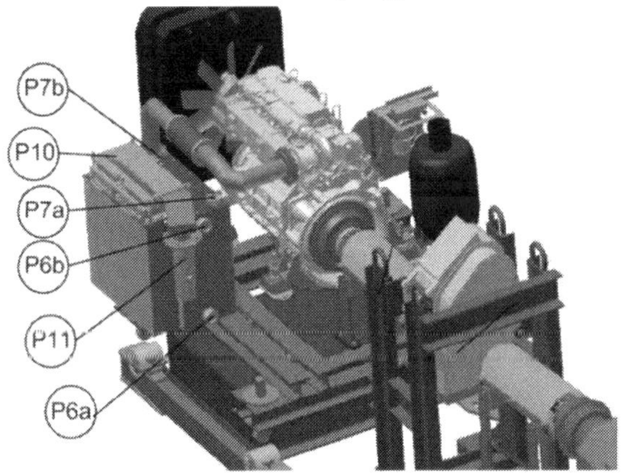

Figure 6. Places of temperature sensors installation.

Due to the fact that the authors intended to compare the results of the pressure free heat exchanger tests with the test results of the currently used wet exhaust gas scrubbers, the testing process was divided into four stages:

Stage1 Exhaust gas tests (temperature measurements) with built-in wet scrubber.

Stage2 Exhaust gas tests (temperature measurements) with built-in PFHE heat exchanger with d=20 pipe system.

Stage3 Exhaust gas tests (temperature measurements) with built-in PFHE heat exchanger with d=44.5 pipe system.

Stage4 Exhaust gas tests (temperature measurements) with built-in PFHE heat exchanger with d=20 pipe system, with partially blocked exhaust gas flow.

It was assumed that the measurements would be taken for the operating points of the diesel engine (engine + engine test bench) as in Table 2.

Table 2. List of operating points of the diesel engine.

Operating point	Revolutions [1/min]	Torque [Nm]	Engine load [kW] *	Operating time [min]
M1	Engine idling (800)	-	-	30
M2	1500	191	30	50
M3	1500	287	45	25
M4	1500	350	55	30
M5	1500	363[1]	57	30
M6	1500	350	55	25
M7	1500	287	45	25

[1] maximum torque

Before starting the tests, each of the tested heat exchangers was filled with water. For a wet scrubber it was 141 l, while for the PFHE it was 103 l. The result of the exhaust gas flow through the heat exchanger was water rushing and pressure increase in it. This phenomenon occurred especially in the case of a wet scrubber. For this reason, to measure the water level correctly, after the end of the measuring cycle at each operating point, the diesel engine was switched off.

The time of conducting measurements at each operating point resulted from a stabilization of temperatures for the first tested configuration (wet scrubber).

5. Measurement results

Temperature measurements in the following points: P1, P3, P5, P8, P10, P11 and P12, are important for a verification of the PFHE correct operation.

In Table 3 the results of maximum temperature measurements in the selected points at testing each version of the device in each operational point are listed.

The symbols are as follows:
MP- wet scrubber of exhaust gases,
PFHE 44.5- pressure free heat exchanger with pipes d=44.5 mm,
PFHE 20- pressure free heat exchanger with pipes d=20 mm,
PFHE 20z- pressure free heat exchanger with pipes d=20 mm,
with partially blocked exhaust gases flow.

Table 3. Maximum temperatures in each mode of operation.

Operating point	Temperature of P1 measuring point [°C]				Temperature of P3 measuring point [°C]			
	MP	PFHE 44,5	PFHE 20	PFHE 20z	MP	PFHE 44,5	PFHE 20	PFHE 20z
M1	101.9	106.0	108.3	107.5	74.8	75.3	78.1	77.7
M2	304.3	312.7	306.6	316.2	229.0	234.3	233.7	243.8
M3	396.7	394.9	398.2	405.7	299.2	302.6	304.8	314.1
M4	462.2	460.6	461.4	470.0	351.5	358.9	363.0	371.9
M5	472.2	478.7	474.8	487.6	363.8	377.8	374.0	390.6
M6	465.5	463.8	470.3	480.3	358.7	368.2	378.0	388.2
M7	406.9	404.3	406.1	414.1	315.4	316.1	321.2	329.0

IOP Conf. Series: Materials Science and Engineering **545** (2019) 012011 doi:10.1088/1757-899X/545/1/012011

Operating point	Temperature of P5 measuring point [°C]				Temperature of P5a measuring point [°C]			
	MP	PFHE 44,5	PFHE 20	PFHE 20z	MP	PFHE 44,5	PFHE 20	PFHE 20z
M1	25.4	25.8	26.3	23.6	-	-	26.5	23.5
M2	55.2	77.4	80.9	82.5	-	-	72.9	75.4
M3	60.9	93.9	98.9	98.3	-	-	89.8	91.5
M4	64.4	98.9	99.3	98.6	-	-	104.1	107.0
M5	64.8	97.5	99.1	98.3	-	-	115.9	130.0
M6	64.0	119.5	115.2	120.3	0.0	0.0	140.2	147.7
M7	64.5	122.3	128.3	135.1	-	138.6	138.6	145.0

Operating point	Temperature of 8a measuring point [°C]				Temperature of 8b measuring point [°C]			
	MP	PFHE 44.5	PFHE 20	PFHE 20z	MP	PFHE 44.5	PFHE 20	PFHE 20z
M1	22.3	22.0	22.9	20.8	22.6	22.4	23.3	20.3
M2	51.8	62.2	54.8	57.6	51.7	67.2	59.2	62.9
M3	55.8	80.1	71.4	72.9	56.9	84.5	77.0	79.0
M4	58.6	99.1	86.1	88.3	60.4	104.1	91.8	95.8
M5	61.1	113.9	95.4	102.1	61.2	120.1	101.3	110.6
M6	66.6	124.4	115.8	120.3	71.7	132.2	122.8	131.1
M7	82.8	126.9	117.3	121.2	90.8	131.2	124.1	131.1

Operating point	Temperature of P10 measuring point [°C]				Temperature of P11 measuring point [°C]			
	MP	PFHE 44.5	PFHE 20	PFHE 20z	MP	PFHE 44.5	PFHE 20	PFHE 20z
M1	24.0	23.9	24.5	22.0	23.4	23.8	24.8	22.1
M2	54.3	61.9	54.9	53.8	54.1	76.2	68.3	71.0
M3	60.3	77.0	71.0	72.3	59.8	94.6	85.3	86.5
M4	64.5	94.7	82.4	86.4	63.1	116.6	100.1	104.4
M5	65.1	105.3	91.0	98.9	63.7	130.7	108.9	121.6
M6	65.9	118.8	111.4	117.5	92.5	145.9	133.1	141.5
M7	84.8	114.1	111.1	116.4	107.3	139.6	133.0	139.6

Operating point	Temperature of P12 measuring point [°C]				Ambient temperature [°C]			
	MP	PFHE 44.5	PFHE 20	PFHE 20z	MP	PFHE 44.5	PFHE 20	PFHE 20z
M1	23.8	23.9	24.4	21.9	19.6	17.5	19.0	18.8
M2	52.9	71.9	64.5	67.0	20.8	19.3	19.4	19.0
M3	58.5	88.4	79.4	80.7	20.7	19.8	19.4	19.7
M4	62.3	108.5	93.0	96.8	20.8	20.5	20.3	19.9
M5	62.9	121.8	101.8	112.2	20.5	20.8	20.3	20.7
M6	85.5	135.8	124.8	131.5	20.6	19.8	20.3	21.2
M7	100.2	130.0	123.5	128.9	20.6	19.2	20.4	21.4

IOP Conf. Series: Materials Science and Engineering **545** (2019) 012011 doi:10.1088/1757-899X/545/1/012011

Table 4. Water level in the heat exchanger at the end of operation in each operating point.

Operating point	Water level in the heat exchanger [cm]			
	MP	PFHE 44,5	PFHE 20	PFHE 20z
Before test	50	45	45	45
M1	37	45	45	45
M2	25.5	42.5	42.5	42.5
M3	22	39	40	39.5
M4	15.5	34	33.5	32.8
M5	1)	27	27.5	26.5
M6	1)	24	22	22
M7	1)	21	18	18

1) – water level below the drain plug (< 10 cm)

In Figures 1-5, temperature changes in the selected points versus diesel engine load are shown. Sample photos, made by thermovision camera, imaging the temperature distribution on the heat exchanger are given in Figure 7.

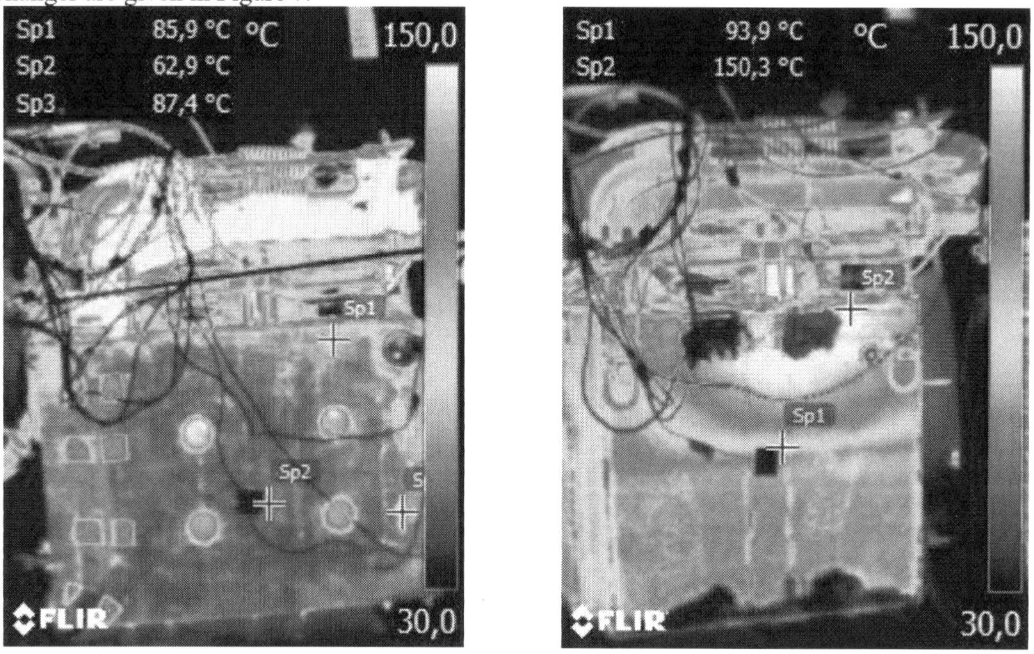

Figure 7. Thermograms of heat exchangers after the cycle completion in M5 operating point. Heat exchanger MP – on the left; pressure free heat exchanger PFHE40 – on the right

As it can be observed from the tests assumptions, relating to the diesel engine operating points, the authors planned to increase load to the diesel engine gradually (Table 2) and then to reduce it. The result of such an action can be observed as changes in temperature at each measurement point.

The temperature of exhaust gases from the exhaust manifold increases with an increase of load to the engine as it is presented in Figure 1. It is understandable, that to keep the assumed revolutions (1500 rpm) at increased torque, more fuel and air has to be fed and then converted to more mechanical energy (torque) and thermal energy which causes an increase of the exhaust gases temperature. Together with a reduction of load to the engine, the temperature of exhaust gases decreases. In Figures 8 and 9, a curve of exhaust gases temperature, when the torque decreases (reduction of load) overlaps the temperature increase curve during the torque increase (load increase). The situation of temperature changes in other selected points of the exhaust system, can be observed e.g. in Figures 10, 11 and 12.

After a reduction of the engine load from the boundary value, the temperature drop curve has a different shape than the temperature increase curve. There is a hysteresis phenomenon. While the temperature of the exhaust gases decreases significantly, when the engine load is reduced, the surface temperature of the exhaust system components decreases more slowly, and in the case of a wet scrubber the temperature of gases increases for the engine load range.

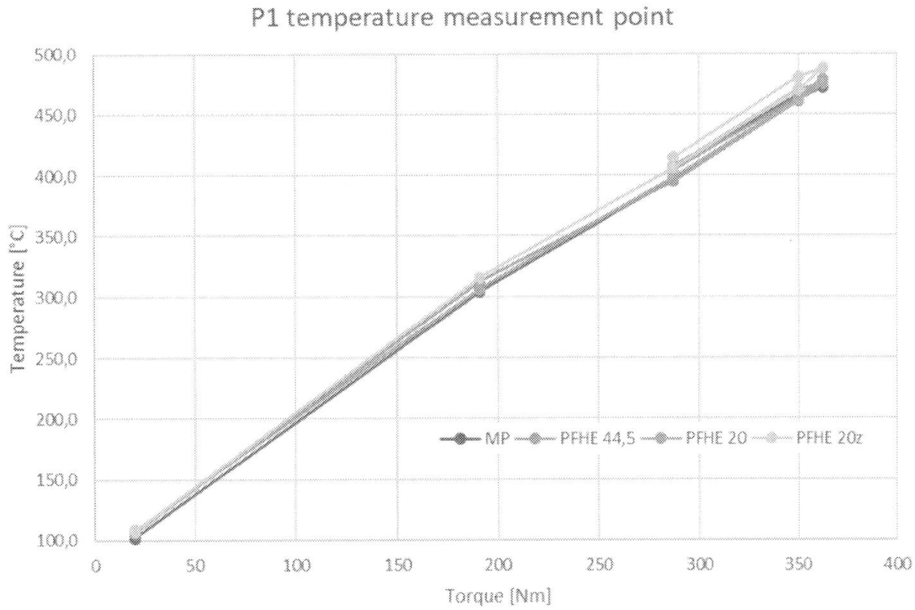

Figure 8. Maximum temperatures of exhaust gases in each operating point in P1 measuring point (turbocharger outlet).

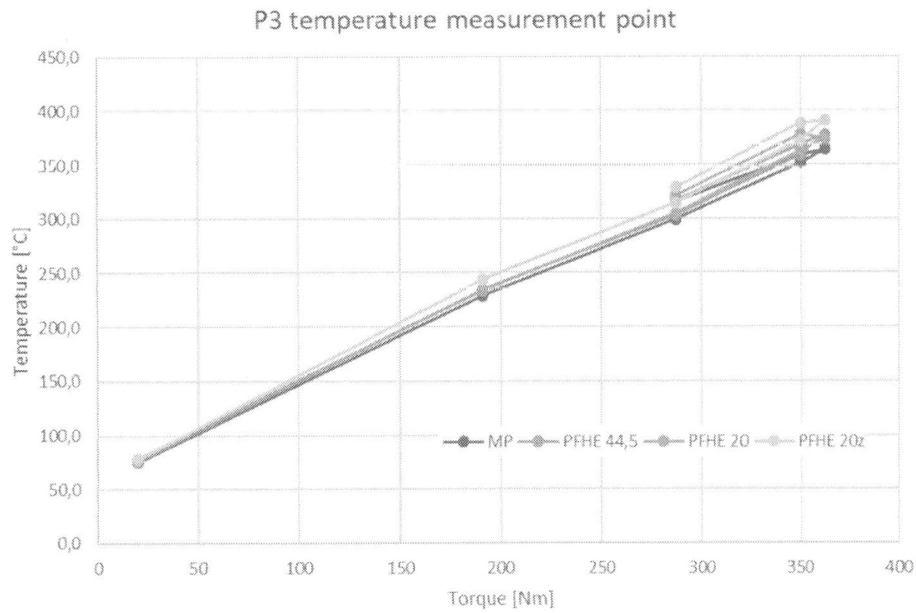

Figure 9. Maximum temperatures of exhaust gases in each operating point in P3 measuring point (exhaust gases temperature on the heat exchanger inlet).

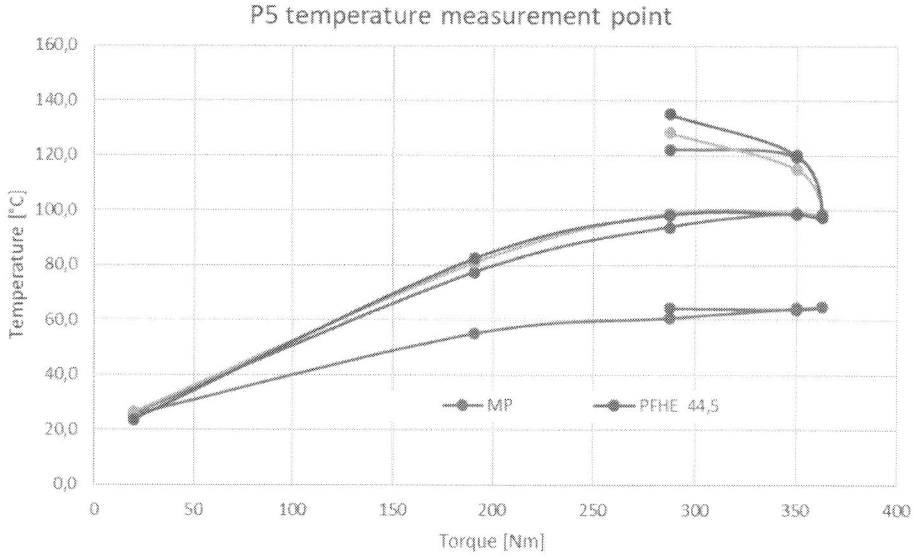

Figure 10. Maximum temperatures of exhaust gases in each operating point in P5 measuring point (external surface of heat exchanger box at half of its height).

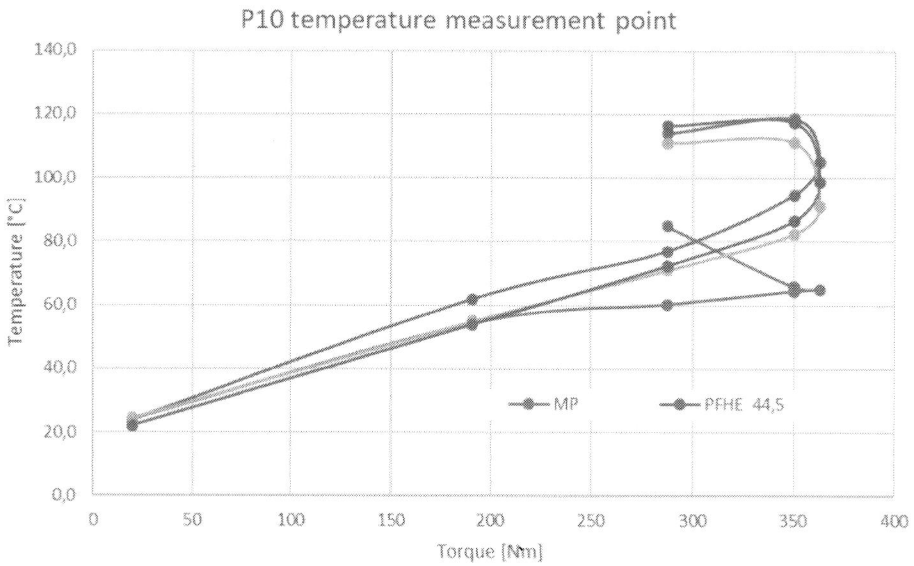

Figure 11. Maximum temperatures of exhaust gases in each operating point in P10 measuring point (surface of exhaust manifold installed on the heat exchanger).

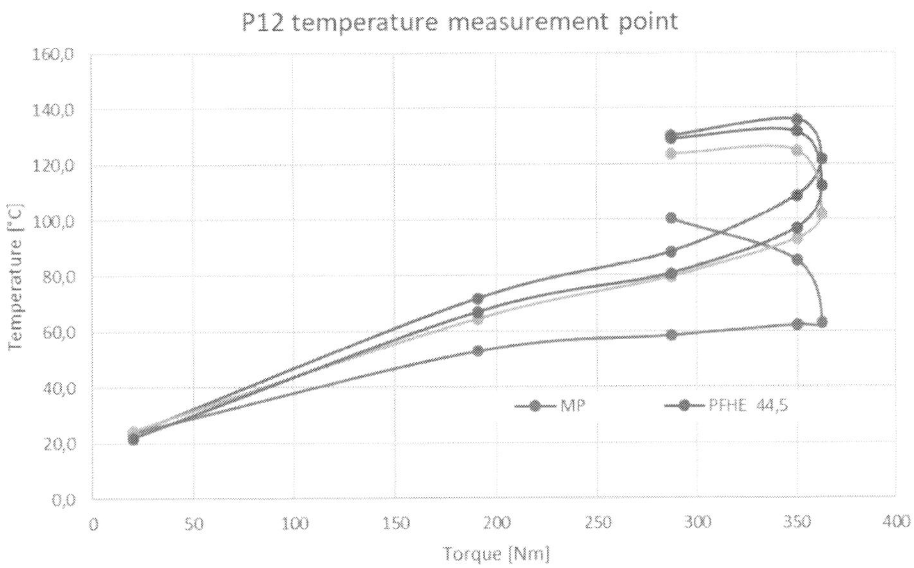

Figure 12. Maximum temperatures in each operating point in P12 measuring point (exhaust gases outlet).

That results from the water loss in the heat exchanger, reducing the efficiency of heat transfer from exhaust gases. Additionally, these systems have the specified heat capacity and the heat, cumulated during the engine operation with the highest load, is emitted to the atmosphere after a reduction in the heat amount generated by a diesel engine. It can be assumed that when the load to the engine is reduced further and after its longer operation, the systems will be cooled and the curves, representing temperature increase and drop connect and the hysteresis loop closes.

6. Conclusions

Testing the PFHE confirmed its high efficiency in cooling the exhaust gases from the engine of maximum power 81 kW. An elimination of water burst during the diesel engine operation as well as significantly lower water consumptions are the main advantages of the PFHE design. A wet scrubber (MP) has a better efficiency, among others, as it is bigger than PFHE – an initial water volume in the wet scrubber is about 1.37 bigger than in the PFHE.

Over the 3-hour engine operational cycle with load bigger than 50% of the maximum torque, in each tested heat exchanger, enabled to cool the exhaust gases below 150°C.

During the tests it was observed that the temperature of the PFHE box external surface was higher than the temperature of the wet scrubber box. However, it happened that the surface temperature of the PFHE exceeded the acceptable temperature of 150°C, but it should be stated that it was an experimental solution. The heat transfer coefficient of the material, used in the tested PFHE, was about 50 in relation to the coefficient 10 of the wet scrubber what explains the surface temperature.

During a normal operation of mine mobile machines, equipped with a diesel drive, a type of a load cycle, as in the test, does not occur [2]. Thus it can be assumed that the PFHE even as a design under-testing should meet the requirements regarding the maximum accepted surface temperature of 150 °C.

After about 3 hours of the engine operation, the wet scrubber was almost completely emptied from water, but the PFHE after 4 hours of engine operation lost 60% of water.

A cooling intensity of the PFHE can be increased by use of additional external cooling system of the cooling system box.

IOP Conf. Series: Materials Science and Engineering **545** (2019) 012011 doi:10.1088/1757-899X/545/1/012011

References

[1] Dobrzaniecki P 2007 *Rozeznanie stanu techniki w zakresie układów wylotowych z wymiennikiem ciepła (suchą płuczką spalin)*. ITG KOMAG Gliwice (not published).
[2] Kaczmarczyk K Brzeżański M *Problemy ekologiczne silników spalinowych eksploatowanych w wyrobiskach podziemnych węgla kamiennego* Monograph nr 46, Gliwice 2015 ISBN 978-83-60708-87-3
[3] Ogrodzki A 1974 *Chłodzenie trakcyjnych silników spalinowych* Wydawnictwo Komunikacji i Łączności Warszawa 1974.
[4] Report No 255/BT/2017 *Badania stanowiskowe bezciśnieniowego wymiennika ciepła ITG KOMAG* Gliwice 2017 (not published).
[5] Suffner H 2018 Chłodzenie spalin w górniczych napędach spalinowych dla kopalń węgla kamiennego - KOMAG *Maszyny Górnicze* **3** pp. 67-80.
[6] Suffner H et al. 2017 *Koncepcja bezciśnieniowego wymiennika ciepła* ITG KOMAG Gliwice 2017 (not published).

IOP Conf. Series: Materials Science and Engineering **545** (2019) 012012 doi:10.1088/1757-899X/545/1/012012

Review of technologies for low-quality solid fuel gasification

A Lutyński[1] and M Lutyński[2]

[1]KOMAG Institute of Mining Technology, Pszczynska 37, 44-101 Gliwice, Poland

[2]Silesian University of Technology, Faculty of Mining and Geology, ul. Akademicka 2, 44-100 Gliwice, Poland

marcin.lutynski@polsl.pl

Abstract. The article discusses World's current status of surface coal and other resources gasification technology. Examples of existing and planned gasification plants efficiencies up to the year 2022 along with the use of syngas for chemical resources production and distribution of plants are given. The SGT (SES Gasification Technology) gasification technology where produced syngas is used for energy production is described.

1. Introduction

Recent agreements within European Union member states impose further reduction of CO_2 emissions. In December 2017 new legislation establishes binding annual greenhouse gas emission targets for Member States for the period 2021–2030. These targets concern emissions from most sectors not included in the EU Emissions Trading System (EU ETS), such as transport, buildings, agriculture and waste. The national targets will collectively deliver a reduction of around 40% in total EU emissions from the sectors covered by 2030 compared with 1990 levels. The Regulation on binding annual emission reductions by Member States from 2021 to 2030 (Effort Sharing Regulation) adopted in 2018 is a part of the Energy Union strategy. It sets national emission reduction targets for 2030 for all Member States, ranging from 0% to 40% from 2005 levels. The national targets are based on Member States' relative wealth. According to the agreement, Poland has to decrease non-ETS emissions by 7%. Additionally, European Commission as part of its proposal for a new Electricity Regulation, the European Commission proposes to introduce an Emission Performance Standard (550 g CO2/kWh) for generation capacity participating in capacity mechanisms. Worth mentioning is the fact that a highly efficient ultra-supercritical coal power plant does not meet the emission standard below 750 g per 1 kWh [1].

Other restrictions of European Commission include decarbonisation, namely phasing out coal as a source of Energy generation. Poland has prolonged these restrictions until 2021. According to the Polish Electricity Association introducing 550 Emission Performance Standard will increase the cost of energy for end-users by approximately 300 million Euros between 2017 and 2040. It will be the consequence of building new power plants fueled by natural gas [2].

One of the solutions of this problem is the gasification technology where solid (coal, biomass, waste) or liquid (oil) fuel is converted into synthesis gas. Synthesis gas is composed mainly of hydrogen and carbon monoxide which are the products of incomplete combustion of fuel. Oxidants in this process are mainly air or oxygen and the choice of the oxidant for the process depends on the fuel reactivity, designation of the end-product and gasifier type. Often, steam is used to regulate the temperature of the process. Molecular disintegration of the feed in the gasifier allows for the separation of impurities such as: sulfur, lead, mercury, etc. It also allows to separate a pure stream of CO_2 which can be further stored or utilized (eg. converted to synthetic fuel).

Gasification technology is known for many years since the first Lurgi gasifier was patented in 1887. Interest in gasification technology has been changing with coal and oil price variations. The higher the price of oil the higher the interest in coal gasification. Various hydrocarbons or biomass could be converted into ammonia, methanol or hydrogen products.

Currently, the development of gasification technology is focused on the production of process gas, which is used in chemical synthesis methods and for electricity generation. According to experts [3,4] the use of gasification in the so-called polygeneration systems, combining the production of electricity and chemical products from coal, mainly motor liquid fuels, methanol or hydrogen is the most attractive option for gasification.

2. Development of gasification technology

Development of gasification technology results in the construction of new installations and gasifiers. Table 1 shows the number of operating, under construction and planned gasification projects and number of gasifiers [5–8]. Figures 1 and 2 are a graphical representation of data given in Table 1.

Table 1. Number of operating, under construction and planned gasification projects and gasifiers from 1999 to 2018 [5–8].

Year	Operating projects/gasifiers	Under construction projects/gasifiers	Planned projects/gasifiers
1999	128/366	n/a	33/48
2001	131/409	n/a	32/59
2004	117/385	n/a	38/66
2007	144/427	n/a	10/34
2010	192/405	11/17	37/76
2013	234/618	61/202	98/550
2016	356/863	108/322	134/685
2017	379/938	131/348	146/734
2018	451/1074	129/355	135/667

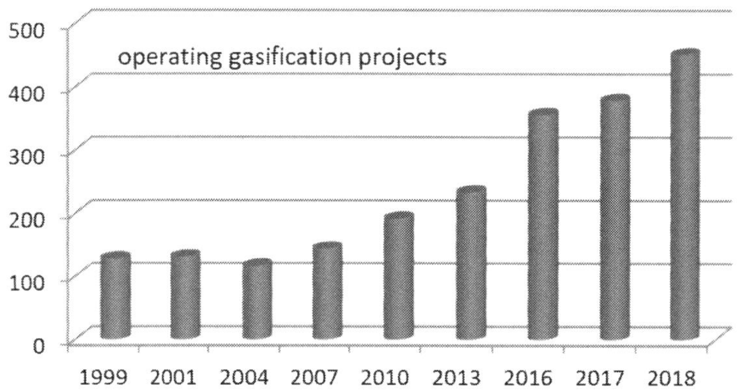

Figure 1. Number of operating gasification projects from 1999 to 2018 [5–8].

IOP Conf. Series: Materials Science and Engineering **545** (2019) 012012 doi:10.1088/1757-899X/545/1/012012

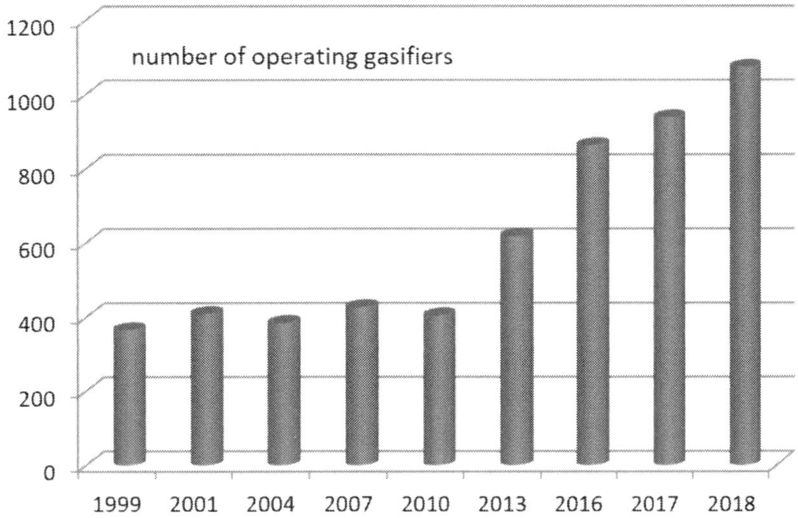

Figure 2. Number of operating gasifiers from 1999 to 2018 [5–8].

Over the last years the cumulative capacity of projects has increased, particularly after 2010 (see Figure 3). Currently, the cumulative capacity is approximately 200 000 MWth of synthesis gas and it should reach 425 000 in 2022.

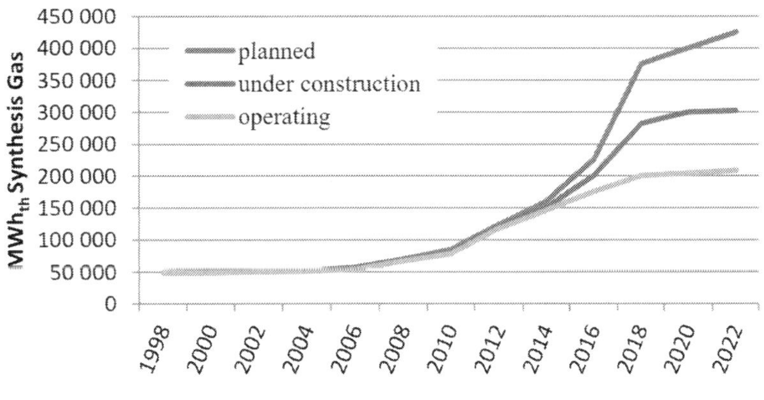

Figure 3. Capacity of operating, under construction and operating gasification project [8].

Synthesis gas is produced from various feedstocks where the most common resource is the hard coal. It is reflected in the number of operating projects and the capacity of installations (Figure 4). Coal has the largest share in the production of synthesis gas among all gasification installations and the number of coal gasifiers had the largest growth in 2017 and 2018. When looking at the number of projects under construction and planned, this trend will continue in 2021 and 2022 [7,8].

IOP Conf. Series: Materials Science and Engineering **545** (2019) 012012 doi:10.1088/1757-899X/545/1/012012

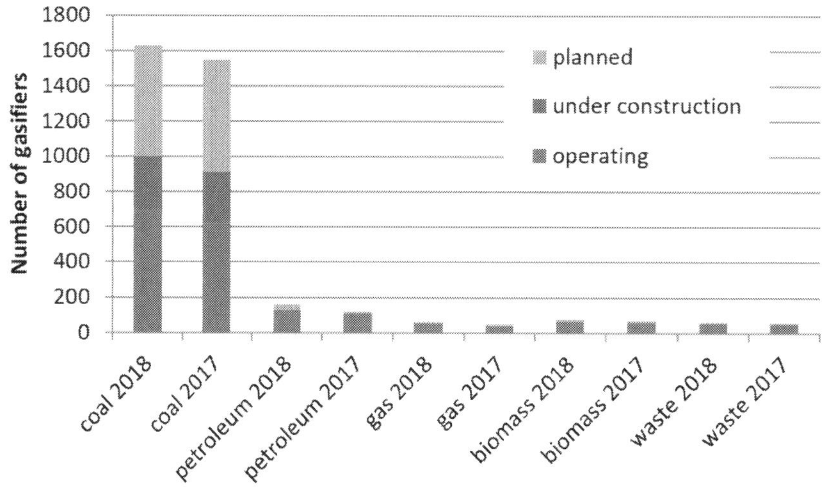

Figure 4. Comparison of operating, under construction and planned gasifiers in 2017 and 2018 by primary feedstock [7,8].

Synthesis gas is used in manufacturing of various products. The majority of it is consumed in petrochemical industry where ammonia, methanol, liquid fuels and hydrogen are produced. It is also used for electricity generation. In Figure 5 consumption of syngas in 2017 and 2018 by application is shown [7,8]. The highest growth in consumption is observed in production of liquid fuels and ammonia.

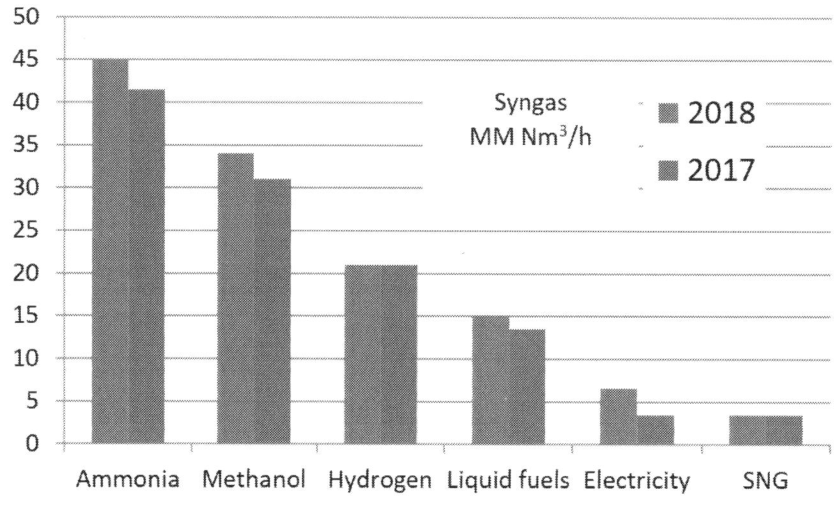

Figure 5. Utilization of synthesis gas depending on application [7,8].

Capacity of operating, in construction and planned gasification installation differs depending on the region where Asia is the region with the largest capacity of installations. The dominance is clear for both operating as well as under construction and planned installations (see Figure 6) [8].

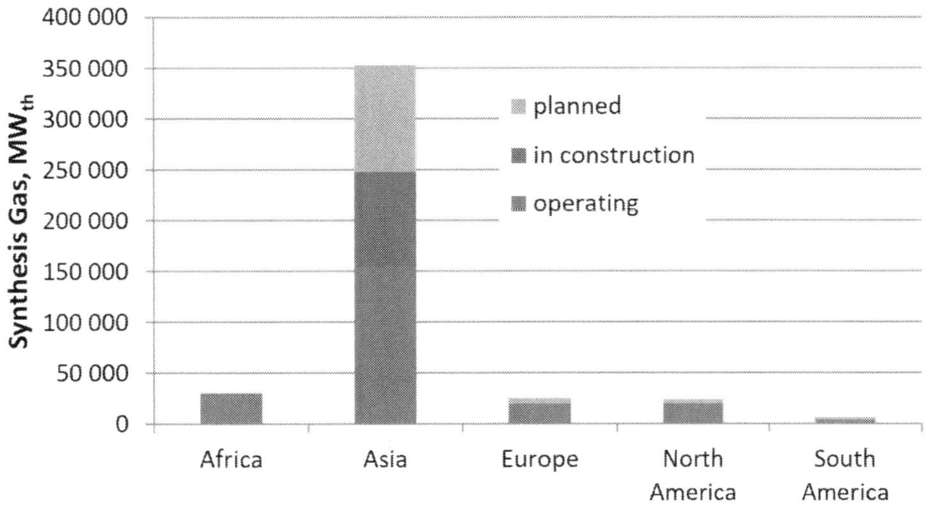

Figure 6. Capacity of installation depending on region in 2018 [8].

3. Forecasts of gasification technology development in Poland

Currently in Poland there is no operating gasification installation of industrial capacity fueled by coal. There are many efforts to develop such installation [2]. One installation is planned near Kedzierzyn-Kozle in southern Poland. According to the estimation it should consume up to 900 000 tons of hard coal per year and cover 15% of national needs for ammonia and methanol. Two large companies: Tauron and Grupa Azoty have signed a letter of intent to erect an installation in 2021-2022 with estimated budget of 400-600 mln Euros. Economic analysis are still being carried out.

Another initiative has been developed between ENEA capital group and coal company BOGDANKA where IGCC plant fueled by coal is planned to be constructed in 2020-2023. Feasibility study is being carried out currently.

EnCoal Gasification Sp. z o.o. is another company that intents to introduce an American SGT - SES Gasification Technology in Poland. Research has been carried out for several years and it appears that this technology would be particularly economically efficient taking into account polish market conditions. Various researchers from Institute for Chemical Processing of Coal, Central Mining Institute and experts from TAURON Wytwarzanie have verified the concept and its application. The SGT technology has been chosen for retrofitting 200 MW boilers in TAURON [9].

On November 9th 2017 American Synthesis Energy Systems Inc. from Houston and polish EnCoal Gasification Sp. z o.o. established Joint Venture named SES EnCoal Energy in Warsaw. This company will have exclusive rights for the SES Gasification Technology (SGT) in Poland [9].

4. Benefits of SGT (SES Gasification Technology)

The SGT is the development of U-GAS® technology which was originally developed by Gasification Research Institute in Chicago. The U-GAS® gasification technology was piloted, demonstrated, and commercially operated on a wide range of feedstocks including bituminous coal, sub-bituminous coal, lignite, biomass, coal char and wastes, and metallurgical coke. As SES began to commercially deploy the U-GAS® technology, the company advanced the technology through modifications that improve fuel utilization, plant efficiency, and cost competitiveness through an expanding intellectual property library – forming SES Gasification Technology. SES is a technological company publicly traded on

NASDAQ since 2007. SES has built and operated several SGT gasifiers in China [9] and has several other gasification projects under development.
Coal feedstock in SGT technology is gasified in innovative fluidized bed reactors under relatively low temperature ie. 840ºC to 1.100ºC. This number is important taking into account the melting point of ashes for polish coals which is slightly above this value. Ash melting may result in reduced air circulation and decreases gasification efficiency. The higher the gasification temperature the more advanced heat protection of gasifiers is required, hence the construction costs are elevated. Pressure in SGT reactor varies from 3 to 40 bars and depends on the designation of synthesis gas. Efficiency of carbon conversion exceeds 85% where conversion into synthesis gas has an efficiency of 99.5%. Remaining 15% of unconverted carbon is converted into heat necessary to keep the process constant. Conventional gasifiers in SGT technology can handle from 200 to 1200 tons of feed per day. Modified gasifiers (under higher pressure) can handle up to 3000 ton per day. Another advantage of this technology is the fuel flexibility, ability to gasify a variety of different coals, biomass or waste. This is also valid for coals where the low quality coals can be used as a feed and this includes coal wastes and slurries (with high content of ash and sulfur). Even low quality coal materials with calorific value below 10 GJ/Mg can be used as a feed. For optimal efficiency a feed of 3500 kcal/kg (14.7 GJ/Mg) [10].

5. Example of SGT technology application

An interesting solution offered by SES company is the I-GAS gasification technology designed for electricity production where gas turbines are fueled by purified synthesis gas. Additional heat generated in the process is used in the steam turbine. Emission of greenhouse gases and other pollutants is comparable to installations fueled by natural gas. In Figures 7 and 8 Solar St Titan 130 by Caterpillar and GE LM2500 by General Electric are presented, respectively.

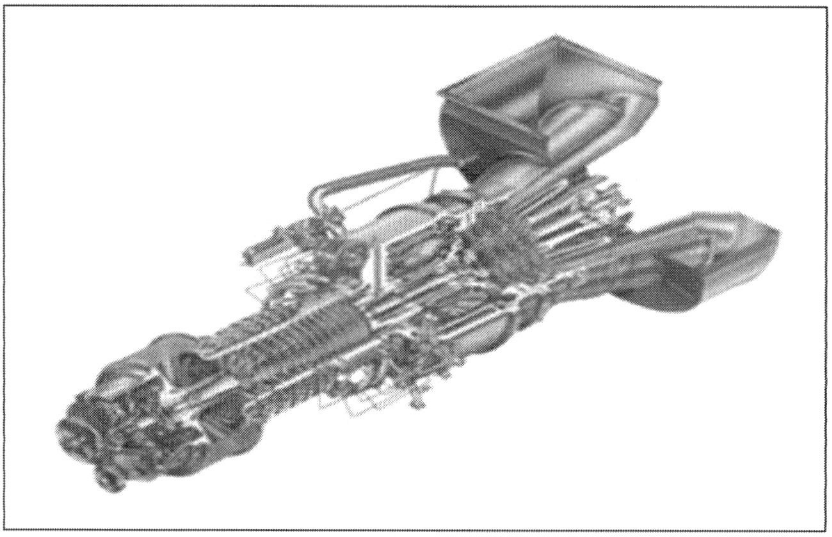

Figure 7. View of Caterpillar Solar St Titan 130 gas turbine [11].

IOP Conf. Series: Materials Science and Engineering **545** (2019) 012012 doi:10.1088/1757-899X/545/1/012012

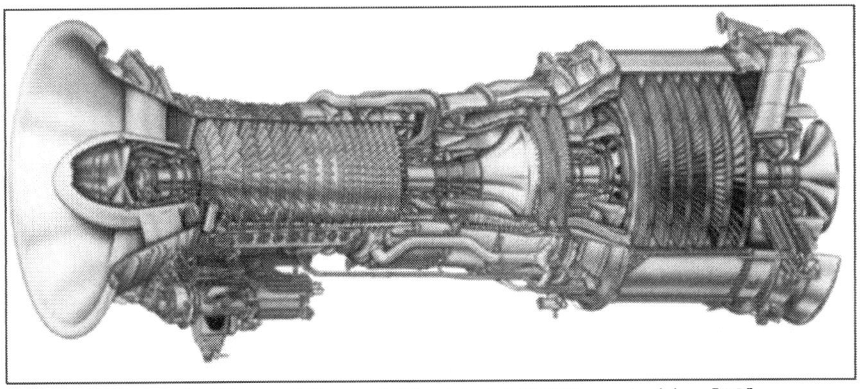

Figure 8. View of General Electric LM2500 gas turbine [12].

Typical configuration of I-GAS gasification installation is presented in Table 2.

Table 2. Typical I-GAS configuration for electricity generation.

I-GAS power capacity, MW	**40**	**80**	**160**
Working pressure, bar	35-50	40-50	40-50
No. of gasifiers, -	1	2	4
No of gas turbines, -	2 St Titan 130	2 GE LM2500	4 GE LM3500
No. of steam turbines, -	1	1	1

6. Conclusions

Abovementioned data indicate increasing interest in the gasification technologies in the World. The number of operating, in construction and planned gasification installation is growing along with the capacity of synthesis gas produced. Such installations can cover energy needs of cities as it commonly observed currently.

In case of Poland where energy mix is mostly based on coal gasification could provide a viable solution to the problem of emission limits. Remaining coal reserves which are relatively abundant could be used without harmful emission penalties. There are already few gasification solutions that could be introduced on the polish market and coal companies express interest in such technologies. Yet, decisions have not been undertaken for any industrial scale installation. Time is running and the following reasons should accelerate gasification on the market:

a) Poland will have to decrease GHG emissions (CO_2) in the transport, housing, agricultural, waste and forestry sector by 7% until the end of 2030 from 2005 levels. Additionally, European Commission proposes to introduce an Emission Performance Standard (550 g CO_2/kWh) for generation capacity participating in capacity mechanisms. These standards could be fulfilled with gasification installation fueled by coal and biomass or waste since part of emissions generated by alternative fuel (RDF – Refused Derived Fuel or biomass) is considered as carbon neutral, hence compliance with 550 g CO_2/kWh standard.

b) SGT technology presented in this article allows gasification of low-quality coals such as coal waste slurries either produced currently or deposited in impoundments with RDF [10]

c) Gasification technology enables separation of pure CO_2 stream which could be further utilized in CCU technologies (Carbon Capture and Utilization). This is a major advantage of gasification where CO_2 separation is much more efficient and economic than in case of conventional coal boilers.

Another breakthrough in the gasification technology is integration of gasifiers with fuels cells. This solution will provide compliance with 550 g CO_2/kWh standard. Extensive research is being carried out in Japan such as EAGLE project coordinated by NEDO (New Energy and Industrial Technology Development Organization) where oxygen-blown coal gasifier is being tested [13].

7. References

[1] Coca M T 2003 Integrated Gasification Combined Cycle Technology: IGCC: Its Actual Applicaton in Spain, ELCOGAS, Puertollano (Elcogás), Editors: Want T, Stiegel G, https://doi.org/10.1016/C2014-0-00849-0

[2] 2017 PKEE: *Bez rynku mocy Polacy zapłacą do 10 mld złotych za niepewność* BiznesAlert.pl

[3] Chmielniak T Sciazko M and Sobolewski A *Nowa Energia* **5-6** 2017

[4] Sciazko M and Chmielniak T 2012 Cost Estimates of Coal Gasification for Chemicals and Motor Fuels Gasification for Practical Applications *IntechOpen* DOI: 10.5772/48556

[5] Higman C 2013 State of the Gasification Industry – the Updated Worldwide Gasification Database *Gasification Technologies Conference* (Colorado Springs, USA)

[6] Higman C 2016 State of the Gasification Industry: Worldwide Gasification and Syngas Databases 2016 Update *Gasification & Syngas Technologies Conference* (Vancouver 19.10.2016)

[7] Higman C 2017 GSTC Syngas Database: 2017 Update *Gasification & Syngas Technologies Conference* (Colorado Springs 17.10.2017)

[8] Higman C 2018 GSTC Syngas Database: 2018 Update *Gasification & Syngas Technologies Conference* (Colorado Springs 29.10.2018)

[9] Lutyński A, Lutynski M. Ksiazkiewicz W, Lada J 2018, ed Klich A, Prostański D *Innowacyjne i przyjazne dla środowiska techniki i technologie przeróbki surowców mineralnych: bezpieczeństwo, jakość, efektywność: KOMEKO 2018* (Gliwice: Instytut Techniki Górniczej KOMAG) pp 116-135

[10] Lutynski A Lutynski M 2017 ed Klich A, Kozieł A, *Innowacyjne i przyjazne dla środowiska techniki i technologie przeróbki surowców mineralnych: bezpieczeństwo, jakość, efektywność: KOMEKO 2017* (Gliwice: Instytut Techniki Górniczej KOMAG) pp 140-156.

[11] www.solarturbines.com. Solar Turbines Powering the Future

[12] www.ge.com

[13] Hiroshi Sano 2016 Development for reduction of CO2 emissions NEDO's Clean Coal Technology (presentation at IEA meeting 6 December 2016)

IOP Conf. Series: Materials Science and Engineering **545** (2019) 012013 doi:10.1088/1757-899X/545/1/012013

THE INFLUENCE OF THE METHODOLOGY FOR SLOPES FORMING IN OPEN PIT MINES ON THEIR STABILITY

T Mzyk and R Jendruś

Silesian University of Technology, Faculty of Mining and Geology, 2 Akademicka Street, 44-100 Gliwice, Poland

E-mail: tadeusz.mzyk@polsl.pl

Abstract. Open pit mines are frequently accumulating significant amounts of material in the form of dumping grounds, landfills or forming land for reclamation. Often the form of emerging dumping grounds is determined by stability analysis of their slopes at the design stage. During the operation of the mining site and the collection of material on the pile, only the geometry of the slope is a subject of control. In many cases, after making slopes of a dozen or so meters height or even up to several tens of meters, and after a certain time has elapsed since their formation, deformation of the escarpments can be observed. At this stage, the only option is to change the geometry, i.e. inclination of a slope or, in the worst case, rebuilding of the dump. In the paper the analysis of the impact of the method of forming slopes and material quality on stability of formed slopes and their safe exploitation has been presented. It also presents a proposal to normalize the methodology of design and construction of slopes in a manner ensuring stability and taking into account the variability of the material parameters from which the slope is to be formed.

1. Introduction

Open pit mines design dumps in such a way that after the storage process, the resulting forms fit into a designed model of reclamation. The slope design takes into account the type and properties of the material being stored, the variability of its physical and mechanical parameters and the assumed geometry of the site (including the height of slopes, safety shelves, land accessibility, etc.). Slides of the slopes of the dumping grounds negatively affect the final form of the terrain, while at the same time causing a large problem in securing the slopes, and ensuring the safety of current mining operations [1]. In order to ensure the final terrain shape, it is necessary in the event of a landslide, taking remedial measures by applying additional safeguards, doing partial or complete reconstruction of the earlier formed slope. This also entails the necessity of incurring additional, high costs of earthworks. For this reason, it is important to identify the causes of landslides in the formed slopes, which supposed to be safe, underwent however a loss of stability.

Pursuant to the definition included in the Act on Environmental Protection Law, ground mass movements are natural, or as a result of human activity, landslides, yielding or removal of surface layers of rock, rock-mantle and soil.

The factors causing the occurrence of slope deformation or the emergence of landslides in mining opencast facilities include, among others [2, 3]:

- the presence of tectonic faults in the bedrock or in the soil of the overburden,
- cracks and relaxation cracks in the ground rocks and in the soil of the overburden,
- leakage of water from lenses and sandy or gravel-stony interbeddings, causing changes in the consistency of cohesive soils,
- large inclination of lithological strata, especially in the edge parts of the deposit in which the slopes and hillsides of the excavation pit are located,
- influence of geodynamic processes (vibrations, oscillations, loads from devices working in the area of the open pit, etc.).

The causes of landslides can be natural or anthropogenic. Natural causes of landslides are the effects of natural forces, such as undercutting of slopes by watercourses, marine abrasion, changes of slope load by sediments (eg growing of rock-mantle covers, silting-up or washing of sediments, aeolian accumulation, etc.), changes in hydrogeological conditions, development of cracks and crevices in the result of the activities of plants and animals, the development of suffusion in the scarp, etc.

Anthropogenic causes of landslides include, above all, undercutting slopes and hillsides, changes in terrain shaping, loading or unloading slopes, dynamic impacts (shocks), changes in land use (deforestation, sodding) or changes in water conditions [5]. The impact of human activity can directly or indirectly lead to the emergence of landslides. The direct impact on the formation of a landslide may have slope undercutting or slope overloading with embankments or building structures [4].

In the case of open pit mines, the most dangerous threats include landslides, especially multi-slope landslides, especially if they occur unexpectedly and in an uncontrolled manner. The emergence of an uncontrolled landslide causes, on the one hand, a threat to people and equipment, and on the other, the need to incur additional expenses to eliminate the effects of its occurrence [4].

For the purpose of analysing the causes of loss of stability in the existing dumping grounds (long-lasting), analysis was conducted on selected three case studies where landslide processes were observed or progressed.

2. Case studies

For the purpose of identifying reasons for the formation of landslides in the slopes of the opencast mine heaps, three cases of slopes of dumps constructed of various materials have been analysed. The analysed slopes are located in one of the excavations of the opencast facility extracting crushed aggregates using blasting works in Upper Silesia region. For the needs of the analysis, the slopes have been marked as slopes A, B and C, and samples of rock material was taken from them for analysis of physical and mechanical properties.

The escarpments are mainly made of material collected from the overburden of the deposit and material produced in the production of crushed aggregate, i.e. clayey sandstone with large amount of coarse rock crumbs (initial unsorted material on slopes A and B) and well sorted material generated in the processing (post-production waste - slope C). On the slopes marked as A and B, material of a similar nature was stored, which did not differ much macroscopically, but came from various stages of the extraction process – from the deposit overburden and from the process of preliminary sorting of spoil [5].

2.1 Case study – slopes A and B

Slopes A and B are made of identical material and with similar technology. There are slopes up to 30 m high, with a safety shelf of 5 meters in the middle of the escarpment (on about 15 meters). They are built from mixed material coming from the overburden of the deposit, mainly clays of various types in the hard-plastic state and material with a similar granulometric composition but originating from the preliminary sorting of the spoil and containing significant admixtures of coarse rock crumbs. On the scarp, the symptoms of mass movement have been noted in the form of slipping off a part of the material and cracks in the ground level in a distance of approx. 5 - 10 m from the edge (figure 1 and 2).

Figure 1. Slides of a part of the material being stored (photo by T. Mzyk).

Figure 2. Cracks in the ground level of the slope (photo by T. Mzyk).

There were also identified places at the foot of the escarpment, in which point outflows of waters with a capacity of several to several dozen litters per minute were visible (figure 3).

Figure 3. Visible water outflows on a scarp of significant intensity (photo by T. Mzyk).

From the scarps A and B samples of the material has been collected for laboratory tests. The samples taken for analysis came from: the foot of the lower scarp (sample A1), the top of the lower scarp (sample A2), from the foot of the upper scarp (at the height of the safety shelf - sample A3) and from the top of the upper scarp (sample A4). In first place, the granulometric composition has been determined (figure 4) and the consistency of the material, then the tests of internal friction angle, cohesion and bulk density have been carried out. The results of the conducted analysis are presented in table 1 and on the graph in figure 4.

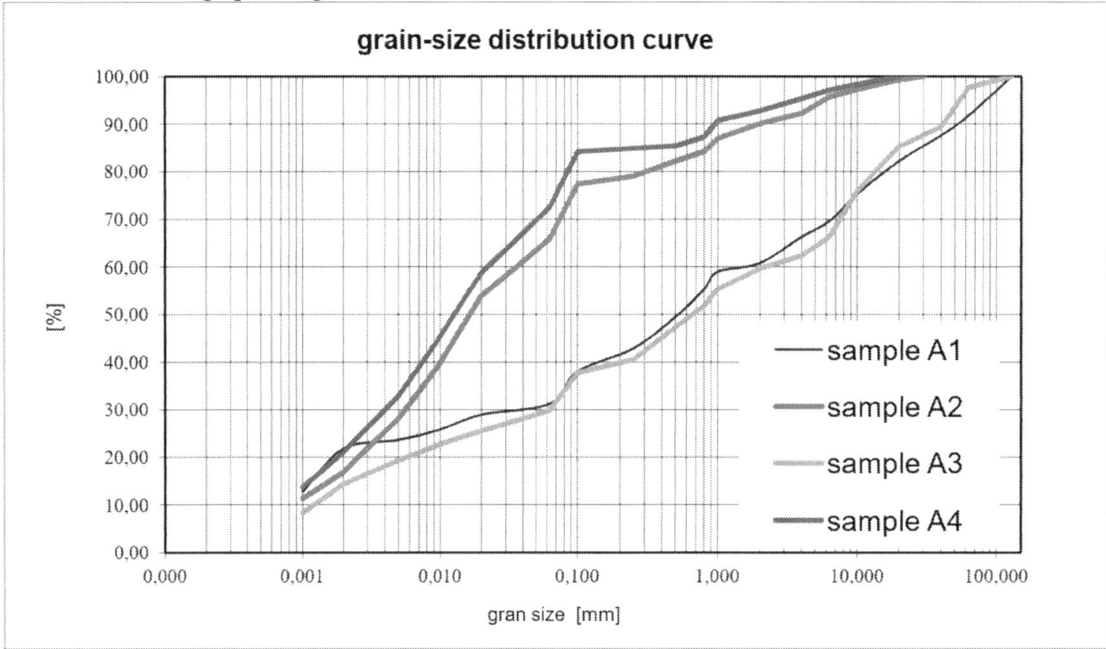

Figure 4. Grain-size distribution curves of analysed samples taken from the slope A.

Table 1. Results of laboratory tests of analyzed samples taken from the slope A.

Sample	A1	A2	A3	A4
Content of fraction (%):				
Stone	8.17	0.00	2.29	0.00
Gravel	30.99	9.89	38.04	7.19
Sand	29.55	24.20	29.82	20.26
Dust	9.52	49.06	15.52	51.59
Clay	21.77	16.85	14.33	20.96
Internal friction angle (^{0})	14.1	16.3	14.6	16.2
Cohesion(kPa)	15.31	19.38	16.29	18.94
Specific gravity (kN/m^3)	20.36	19.74	20.51	19.58
Degree ofplasticity (-)	0.24	0.12	0.22	0.13

The conducted analysis exhibited that the material collected on the slope shows great diversity both in terms of grain-size and variability of physical and mechanical parameters. It can be seen that there is a certain regularity in the parameter change. It was found that the material occurring in the lower part of the escarpment is characterized by an increased content of stony and gravel fraction. The consistency of this material also decreases, i.e. the material increases its plasticity. The identified relationships were tried to be associated with the time duration and technology of slope formation. On this basis, it was found that there is a relationship between slope formation technology and graining.

The increase in the content of coarse fractions in the lower part of the escarpments is due to the fact that the material stored on the pile was transported by trucks and formed by bulldozing it towards the slope [6]. Such technology caused that thick rock crumbs rolled to the foot and formed a layer with a coarser grain size. Such a layer was characterized by the presence of an increased amount of empty voids, which created favourable conditions for the accumulation of precipitation water. Hence the identified outflows of waters with a relatively high flow intensity at the bottom of the escarpment. The accumulation within the lower part of the escarpment, in the layers forming the slope of precipitation water and the presence of voids promotes infiltration of rainwater (probably of considerable range) and influences the accumulated material gradually plasticizing it. Such a situation may in the future cause the intensification of landslide phenomena reasoned by the decrease of physical and mechanical properties of material collected on the slope and within the dumping ground [5].

2.2. Case study – slope C

Slope C is made of well sorted material coming from preparation processes, practically free of fraction with a diameter greater than 20 mm and with a reduced content of clay fraction. This material in dry condition corresponds to loose soils, exhobots little consistency. Slope C is formed to a height of about 30 m, originally with a safety shelf of 5 meters in the middle of the escarpment (at about 15 meters), but in the current state virtually the entire width of the shelf is occupied by slided or rinsed material. On the scarp, the symptoms of mass movement in the form of fissures, cracks, falling off of material, and thresholds in the slope foot as well as large erosive e dilutions (figures 5, 6 and 7) were noticed.

Figure 5. Cracks in the slope foot (photo T. Mzyk).

Figure 6. Thresholds indicating the landslide of the material accumulated on the slope (photo by T. Mzyk).

IOP Conf. Series: Materials Science and Engineering **545** (2019) 012013 doi:10.1088/1757-899X/545/1/012013

Figure 7. Visible deep erosive dilutions on the slope (photo by T. Mzyk).

Due to the diversified humidity and permeability of the material collected on the slope C, samples were taken from places of poorly permeable nature (visible stagnant precipitation water basins) and material with good permeability. The samples taken for analysis came from: the foot of the lower escarpment (sample C1), the top of the of the lower escarpment (sample C2), from the foot of the upper slope (at the height of the safety shelf - sample C3) and from the top of the upper escarpment (sample C4). First, the granulometric composition was determined (figure 8) and the consistency of the material, then the tests of internal friction angle, cohesion and bulk density have been carried out. The results of the conducted analysis are presented in table 2 and on the graph in figure 8.

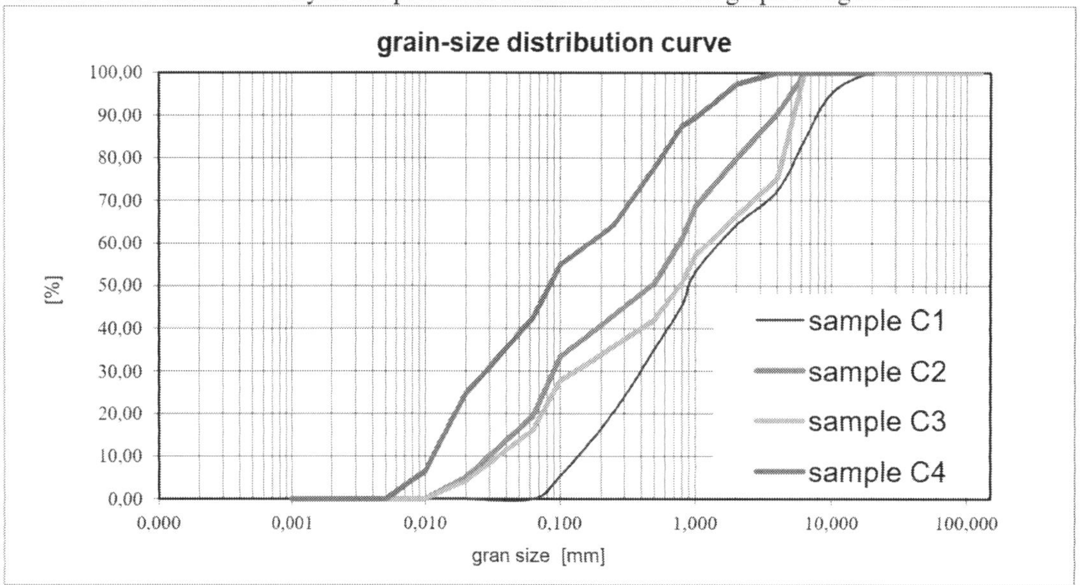

Figure 8. Grain-size distribution curves of analysed samples taken from the slope C.

IOP Conf. Series: Materials Science and Engineering **545** (2019) 012013 doi:10.1088/1757-899X/545/1/012013

Table 2. Results of laboratory tests of analyzed samples taken from the slope C.

Sample	A1	A2	A3	A4
Content of fraction (%):				
Stone	0.00	0.00	0.00	0.00
Gravel	35.73	20.09	33.55	2.62
Sand	64.27	60.54	50.35	54.90
Dust	0.00	19.37	16.11	42.48
Clay	0.00	0.00	0.00	0.00
Internal friction angle (^{0})	37.4	36.8	38.1	35.4
Cohesion(kPa)	6.12	0.00	3.41	0.00
Specific gravity (kN/m^3)	22.74	21.98	22.37	21.63

The conducted analysis showed that the material collected on the slope exhibits little differentiation in terms of grain-size distribution and variability of physical and mechanical parameters. It can be seen that there is a certain regularity in the parameter change. It was found that the material occurring in the lower part of the slope is characterized by slightly higher specific gravity values, a slightly higher value of the internal friction angle, and shows minimal consistency values. This material can be classified as non-cohesive one. The identified relationships were tried to be associated with the time duration and technology of slope formation. On this basis, it was found that there is a relationship between slope formation technology and graining. The material accumulated at the edge of the slope is much less densely compressed and easier to rinse. In particular, at the edges of the escarpment, precipitation water runoff involves leaching significant amounts of material and accumulating them in the lower parts of the escarpment, hence virtually the disappearance of the security shelf, buried by the rinsed material from the above-mentioned parts of the escarpment. The same situation takes place in the lower part of the escarpment where it also accumulates leached material. In general, this reduces the slope angle [5].

3. Analysis of slope stability

One of the three calculation approaches introduced by PN-EN 1997-1 was used to assess stability. In Poland, in accordance with the national annex, approach No. 3 is used for stability assessment. Partial coefficients have been divided into three groups and take into account the coefficients used for interactions or their effects (A), coefficients used for ground parameters (M) and coefficients used for resistances occurring on the slip surface (R) [7, 8].

In the DA3 approach, for the analysis of slopes stability and overall stability of the impact on the ground foundation (structures impacts, traffic load) is considered as a geotechnical impact and a set of A2 load coefficients is applied, i.e. factors are taken into account for variable impacts. The DA3 also includes coefficients for strength parameters [7, 8].

On the basis of determination of actual parameters obtained from laboratory tests of samples of the material dumped on scarps, a slope stability analysis was conducted to answer the question: whether the formed slopes are stable and possibly what additional factors may influence initiation of a landslide. Stability analysis was carried out using the Fellenius, Bishop and Morgenstern Price method in the Geo Slope 1.12XE program. An automatic determination of the center of rotation of the critical slip plane was adopted. Next, attempts have been made to find a slip surface with the characteristics (course, range) identical as observed on the escarpment [5].

4. Identification of the causes of a landslide on the analysed slopes

According to the requirements of EUROKOD 7 [7, 8] for slopes loaded in the foot and endangered by mass movements, the minimum values of the stability index should be increased in the range from 1.5 to 2.0.

IOP Conf. Series: Materials Science and Engineering **545** (2019) 012013 doi:10.1088/1757-899X/545/1/012013

The slopes in the analysed excavation were and still are subjected to shocks and vibrations resulting from the used method of mining operations. For this reason, the values of the stability index above 2.0 obtained from modelling - stability analysis, allow to recognize such slopes as safe – stable. Smaller values indicate that the slope of designed geometry made of a specific material has been designed incorrectly.

On the basis of the results of the slope analysis, it was found that the minimum value of the stability index FS for slope A is above 2.35 and for slope B is 3.17. The obtained values indicate that the escarpments should be stable and the causes of landslides within them belong to other factors, not considered in the process of designing. Previously mentioned, highly probable factors related to the separation of material during its dumping on the escarpment and leaving the voids in the ground, in which the fine material is probably currently being washed in, causes the formation of depressions and zones with significantly reduced physical and mechanical properties. In the case when these zones are located in the area of the slope, it is in these areas that the material tends to slip. In addition, the dynamic impacts contribute the phenomenon of thixotropy causing even temporary plasticization of the material within the escarpment [5].

The FS stability factor value for slope C is 1.27. The potential slip plane with the lowest values obtained from calculations, covers the part of the slope without giving rise to the deformations observed in the actual slopes. The values of the obtained stability factors indicate that the A and B slopes are stable. On the other hand, taking into account the low value of the stability factor C, it should be recognized that the cause of the formation of a landslide within it, is a design error, not taking into account the characteristics of the material built into it. This is confirmed by erosive dilution and successive covering of material on the embankment [5].

5. Summary and Conclusions

The conducted analysis made it possible to assess the slope stability at an open pit mine facility and to identify the causes of the formation of landslides within the slopes of the dumping grounds of this plant. It was found that in one case the loss of slope stability results from its incorrect design. In the other two cases, the cause of the formation of a landslide in the slope formation technology was identified. It was also determined that the effects of the method of forming slopes will be significantly extended over time and in the future may cause the landslide process to intensify.

Conclusions for the practice resulting directly from the conducted analysis can be formulated as follows: in determining the safe inclination of slopes one should take into consideration:

- the type of land in which the slope will be made, - the controlled process, soil or material collected should be properly formed or modified to the designed parameters,
- in particular, the size of expected loads, vibrations, impacts in the vicinity of the slope, the possibility of changing parameters as a result of irrigation of collected material and as a result of dynamic impacts should be taken into account,
- it is necessary to precisely determine the expected lifetime of slopes and their durability - scarps that are the final slopes have to take into account the manner of the final use of the area,
- effects of a possible loss of slope stability, the possibility of securing, remodelling, changing the way of use, should be included in the method of shaping, including taking into account the possibility of changing the properties of the material (modification of its properties),
- at the design stage, possible ways of strengthening, securing the slope should be taken into account (perform reverse filters on a regular basis, apply geosynthetics, anti-vibration barriers).

Scarps made in open pit mining plants should be designed and executed as building objects with strict adherence to technology and monitoring principles at each stage of their execution. Abandoning such an approach results in very expensive works related to the transfer of significant quantities of material or the use of expensive technologies to protect landslides.

IOP Conf. Series: Materials Science and Engineering **545** (2019) 012013 doi:10.1088/1757-899X/545/1/012013

6. References

[1] Mzyk T, Gorol M and Czajkowska A 2016 Zastosowanie odbiornika GNSS do monitoringu wpływu działalności górniczej na powierzchnię terenu *Systemy wspomagania w inżynierii produkcji* **5**(17), ed Pozzi M Publishing house PANOVA pp 176 – 190
[2] Szruba M 2016 Stabilizacja skarp i zboczy *Nowoczesne Budownictwo Inżynieryjne* **5**, pp 74–78
[3] Instrukcja ITB Nr 424/2011 Ocena stateczności skarp i zboczy. Zasady wyboru Zabezpieczeń (Guide of Institute Building Technology)
[4] Mzyk T and Gorol M 2017 Ocena przydatności terenów pogórniczych do celów budowlanych w rejonach dawnej płytkiej eksploatacji górniczej *Systemy wspomagania w inżynierii produkcji* **6**(3) ed Pozzi M Publishing house PANOVA pp 160 – 170
[5] Mzyk T 2018 Ekspertyza w zakresie oceny stateczności skarp zwałowiska wewnętrznego. Strzelce Opolskie *(not published)*
[6] Generowicz A, Gaska K and Hajduga G 2018 Multi-criteria Analysis of the Waste Management System in a Metropolitan Area *E3S Web of Conferences* **44** p 00043
[7] PN-EN 1997-1:2004 Eurocode 7 – Geotechnical design – Part 1:General rules
[8] PN-EN 1997-2: 2007. Eurocode 7 – Geotechnical design – Part 2: Exploration and testing of ground foundation

IOP Conf. Series: Materials Science and Engineering **545** (2019) 012014 doi:10.1088/1757-899X/545/1/012014

Underground coal mine workings as potential places for Compressed Air Energy Storage

M Lutyński[1] Ł Bartela[2] G Smolnik[1] and S Waniczek[3]

[1]Faculty of Mining and Geology, Silesian University of Technology, Akademicka 2, 44-100 Gliwice, Poland

[2] Faculty of Energy and Environmental Engineering, Konarskiego 18, 44-100 Gliwice, Poland

[3]Energoprojekt-Katowice SA, ul. Jesionowa 15, 40-159 Katowice, Poland

marcin.lutynski@polsl.pl

Abstract. The article gives a brief overview of current developments and projects of Compressed Air Energy Storage (CAES). Typical CAES configurations such as Adiabatic CAES and Diabatic CAES are described. The concept of air storage in isolated workings of closed coal mine is presented taking into account availability of such places in the Silesian Coal Basin of southern Poland. The article also discusses major challenges of such concept such as insulation of underground workings, geomechanical stability of workings and site availability. As a proof of concept examples of underground coal mines converted into natural gas storage sites are given. Types of underground workings that could serve as a part of potential compressed storage site are listed and an example of volume calculation available in coal mine for storage is given.

1. Introduction

The key element providing flexibility in energy generation and supporting the use of renewable energy resources is energy storage. Energy storage serves various purposes where the primary one is to balance centralized and distributed energy generation, particularly in the era of renewables, while the other is providing energy security. The share of renewable energy sources (RES) for electricity production in European Union is constantly growing with projected targets reaching at least 32% in 2030 and possibly over 80% in 2050[1]. In USA 17% of all electricity generation is supplied by RES, China has also set targets to reduce its carbon emissions per unit of gross domestic product by 60–65% by 2030 from the 2005 levels where renewables will play a crucial role reaching 20% by 2030[2]. This means that significant amounts of excess renewable energy (on the order of TWh) will start to emerge in EU and in other large economies, with surpluses characterized by periods of high power output (GW) far-off in excess of demand. It may happen that this periods will coincide with periods of low RES energy generation, hence non-renewable generation capacity will be required. This will also put strain on the transmission and distribution system, balancing supply and demand of electricity will become even more difficult. An extensive study on energy storage in EU [3] revealed that demand for power-to-power storage will grow up to 10 times the currently installed capacity, or about 400 GW in the EU. There is no doubt that the problem of energy storage will be a challenge for energy sector in the next decades. This, in combination with limited number of solutions for large energy storage systems may cause problems for the entire EU energy system.

To date, two main large energy system storage are in use. First is the Pumped Hydroelectric Storage (PHS). PHS is an energy storage technology with a long history, numerous installations and reliable operation history. It also has the highest energy capacity. With an installed capacity of 127–129 GW in 2012, PHS represents more than 99% of worldwide bulk storage capacity and contributes to about 3% of global generation. Another solution is the Compressed Air Energy Storage (CAES) which can provide power output of over 100 MW with a single unit [4]. The PHS installations require convenient topography and vicinity of a lake or a river. In case of CAES the most important issue is the geology setting namely, existence of leak tight and pressure resistant structures which can be easily accessible with boreholes. So far, only leached salt domes proved to fulfil these conditions. Other concepts of energy storage which currently are at different levels of technology readiness are shown in Fig. 1.

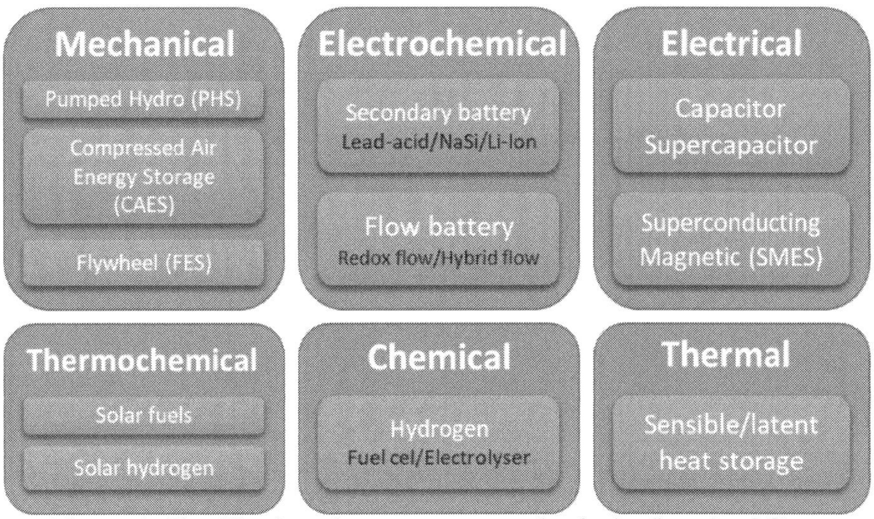

Figure 1. Classification of energy storage technologies (based on[4])

In this paper, a brief review of current developments of CAES technology is given and the concept of storing compressed air in abandoned or closed coal mines is described with particular focus on the Upper Silesian Coal Basin region in southern Poland.

2. The concept of Compressed Air Energy Storage

The general concept of Compressed Air Energy Storage is fairly simple. Excess energy runs compressors which pump air into underground cavern and the air is decompressed when the energy is needed through a turbine which is fueled by natural gas. This configuration is called diabatic CAES (see Figure 2). The major problem in conventional diabatic CAES is the low efficiency as a result of heat loss during the compression stage and energy used for heating the air before the turbine. Other concepts include Adiabatic CAES (A-CAES) presented in Figure 3 where heat is recovered and stored and isothermal CAES (I-CAES) where increase and decrease of temperature is avoided when air is compressed and decompressed respectively (see Fig 4).

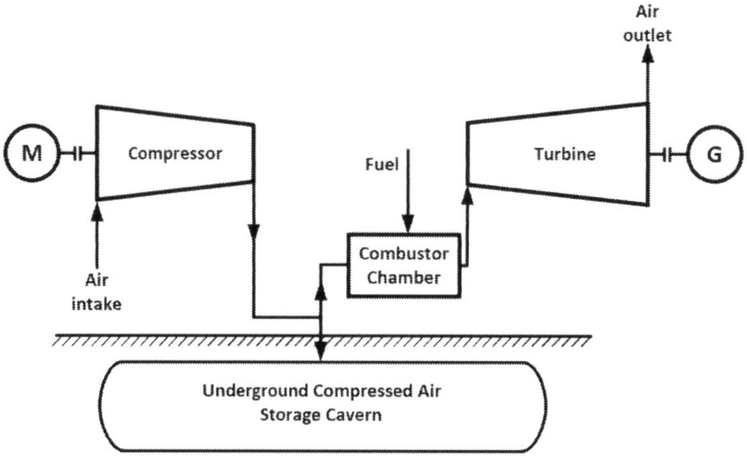

Figure 2. Concept of diabatic CAES (D-CAES)

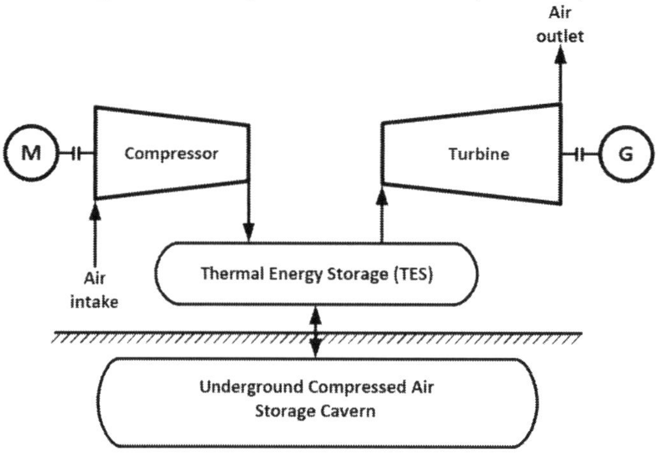

Figure 3. Concept of adiabatic CAES (A-CAES)

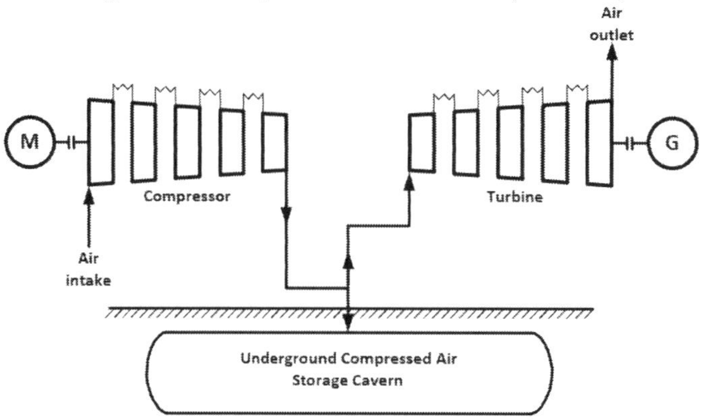

Figure 4. Concept of isothermal CAES (I-CAES)

Two D-CAES plants are in operation today. The first one is the 321 MW Huntorf plant in northern-west Germany belonging to E.ON Kraftwerke. For air storage underground salt cavern is used. In case of Huntorf there are two salt caverns of 310 000 m^3 volume at a depth of 600m with a pressure tolerance between 5.0 –7.0 MPa. Under working condition the plant runs in a daily cycle with eight hours of compressed air charging and two hours of expansion operation at a rated power of 290 $MWTh_e$. The second plant is the McIntosh of 110 MW capacity located in Alabama, USA. It belongs to PowerSouth company and was designed to be operated for up to 26 hours continuously at its full power. The storage capacity of a single cavern is approximately 560 000 m^3 at a depth of 450 m and stores compressed air in the range of 4.5 MPa to 7.4 MPa [5].

Other concepts under development include the ADELE project coordinated by RWE Power from Germany. In this concept, heat generated during the compression stage is stored in a separate Thermal Energy Storage (TES) unit. The TES unit will be constructed on the surface and consists of a solid material (concrete, stone) which will accumulate heat (see, Figure 5). During the discharge phase heat will be recovered from the TES unit. The A-CAES will be equipped with approx. 240 MW compressor power, 260 MW turbine power and approx. 1- 2 GWh capacity (i.e. 4 to 8 full load hours). The pressure of stored air will reach 7 MPa and the temperature of storage in TES unit 600-650°C. It is anticipated that the efficiency of this CAES system will reach 70% [6].

Figure 5. Laboratory testing of concrete reinforcement in the Thermal Energy Storage unit of ADELE project [6].

Another concept is the Norton project which is under development by FirstEnergy Corporation from USA. In this concept an underground limestone mine will be converted into compressed air storage. The Norton Mine will be developed for air storage by constructing conventional gas-type wells into the mine for air injection and withdrawal. The mine is currently open to the atmosphere only through two existing mine access shafts. The mineshafts will be sealed with impermeable plug materials. The mine was developed laterally with the use of room-and-pillar method of extraction. The depth of workings is approximately 670 meters below the surface. Mine stability has been proven for over 60 years and in-depth studies revealed that mine is stable as a pressure vessel [7].

3. Efficiency of CAES

The primary indicator of the thermodynamic effectiveness of each energy storage system is the energy storage efficiency. This indicator is defined as the quotient of the amount of energy obtained at the stage of discharge of the system and the amount of energy directed to the system at the charge stage. For an adiabatic or isothermal CAES system, the definition can be written as follows:

$$\eta_{\mathrm{CAES}} = \frac{E_{\mathrm{elG}}}{E_{\mathrm{elC}}}, \tag{1}$$

where: E_{elG} - the electricity fed into the grid, E_{elC} - the electricity consumed by the compressor.

In the case of diabatic CAES, the use of the definition (1) may not be sufficient. Here, at the discharging stage, there is an additional flow of energy through the balance cover of system. It is the chemical energy of gaseous fuel, the oxidation of which in the part of the volume of air directed from the storage tank makes it possible to significantly increase the thermodynamic potential of the gas at the expansion stage.

The literature on the subject uses different definitions of efficiency for the assessment of diabatic systems. Most often, in the case of analyzes of the most popular systems operating in the USA and Germany, the following definition is used:

$$\eta_{\mathrm{D-CAES(1)}} = \frac{E_{\mathrm{elG}}}{E_{\mathrm{elC}}+Q_{\mathrm{f}}}, \tag{2}$$

where: Q_{f} – the chemical energy contained in supplied fuel.

However, the use of equation (2) for determining efficiency can be controversial due to summation in the denominator of various forms of energy, i.e. electricity and chemical energy of fuel. In [8] the authors discuss the use of three alternative definitions of the efficiency of diabatic systems, which from a thermodynamic point of view are less controversial:

$$\eta_{\mathrm{D-CAES(2)}} = \frac{E_{\mathrm{elG}}}{\frac{E_{\mathrm{elC}}}{\eta_{\mathrm{elR}}\cdot\eta_{\mathrm{tr}}}+Q_{\mathrm{f}}}, \tag{3}$$

$$\eta_{\mathrm{D-CAES(3)}} = \frac{E_{\mathrm{elG}}-E_{\mathrm{elC}}}{Q_{\mathrm{f}}}, \tag{4}$$

$$\eta_{\mathrm{D-CAES(4)}} = \frac{E_{\mathrm{elG}}-Q_{\mathrm{f}}\cdot\eta_{\mathrm{gas}}}{E_{\mathrm{elC}}}. \tag{5}$$

where: η_{elR} - the efficiency of a reference base load power plant, η_{tr} – the efficiency of electricity transmission to the energy storage system, η_{gas} - the efficiency of conversion of chemical energy to electricity.

The use of one of the definitions (2) - (5) makes it more convenient when comparing diabatic CAES system with other energy storage technologies.

4. Examples of mines converted into gas storage

As it was mentioned in the previous chapters, the technology of storing compressed air in salt caverns has been proven by successful operation of Huntorf and McIntosh plants for over 30 years. Nevertheless, the availability of salt structures for CAES is limited [9]. One of the place where compressed air could be stored are underground mines. This concept has been presented in [10,11]. Abandoned or closed coal mines have been used for natural gas storage in the past. Three industrial scale installations have been in operation for several years. Two coal mines were used as temporary natural gas storage in Belgium: Peronnes and Anderlues, located between Mons and Charleroi and a third mine was in Leyden, USA. In contrary to compressed air, methane is easily adsorbed on coal and the storage capacity of these mines was largely enhanced by sorption. The Anderlues storage facility was operating between 1980 and 2000. This site had a relatively low storage pressure (max 0.35 MPa) but the reservoir volume in total was estimated between 6 and 10 Mm^3. This was enough to store approximately 20 Mm^3 of CH_4. Peronnes Mine was also located in the gassy Hainaut Coalfield in Southern Belgium and could store 120 Mm^3 of CH_4 [12]. Another good example is the Leyden mine in Colorado, USA where storage operation started in 1961. Although there was a public debate on the possible leakages from the site it was proven to be successful for over 40 years. The gas was stored under the maximum operating pressure of 1.8 MPa with maximum storage capacity of 21 Mm^3 [13]. Cross section of Leyden mine is shown in Figure 6.

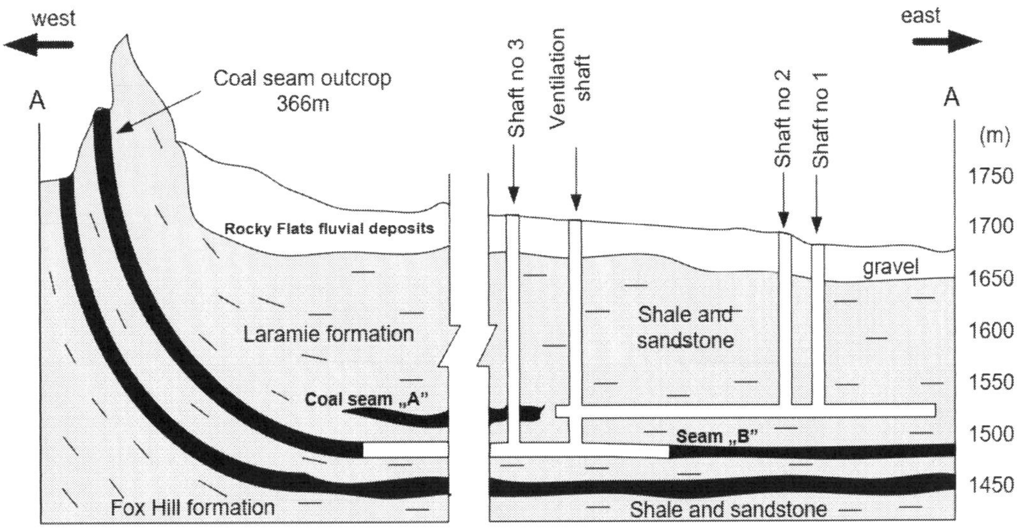

Figure 6. Cross-section of Leyden mine [13].

Concept of storing natural gas in coal mine was also explored in the Lower Silesia Coal Basin in Poland where Nowa Ruda coal mine was planned to be converted into underground gas storage. Estimated storage capacity varied depending on the storage pressure and storage mode. If only caverns, porous rocks and other workings (without remaining coal) were taken into account the storage capacity would vary from ca. 19 Mm^3 at the working pressure of 0.25 MPa to over 50 Mm^3 for working pressure of 0.6 MPa [14]. Storage capacity could be largely enhanced by coal adsorption to over 70 Mm^3

5. Underground coal mine workings as underground compressed air storage

Using underground coal mine workings as a potential compressed air storage seems to be a questionable concept at a first glance. However, when taking into consideration potential benefits it could be a viable and economic solution. This concept has many benefits and these are:

- relatively high storage capacity when comparing to salt caverns,
- convenient location in urbanized areas where energy storage is an often problem,
- already available geological survey data,
- accessibility to mine maps,
- well recognized hydrogeology of the potential site,
- potential access to state/regional or other funds for mine reclamation which could be used to develop the concept,
- lower cost in comparison to other underground sites since the excavations already exist and existing roof/wall support could be used,
- safe operation and protection against external factors (weather, equipment failure, terrorist attacks) since the underground system is separated from the surface and connected only by valves and pipes,
- large surface area available for the plant (former mine surface facilities),

This concept would be a viable solution only if the drifts and shafts are properly sealed and separated from the remaining coal seams. Another issue is the geomechanical stability of overburden in the longwall mining with roof caving method. Currently, most of the modern coal mines use a longwall method without backfilling. As a consequence, roof collapses and overlaying rock layers may fracture (Figure 7).

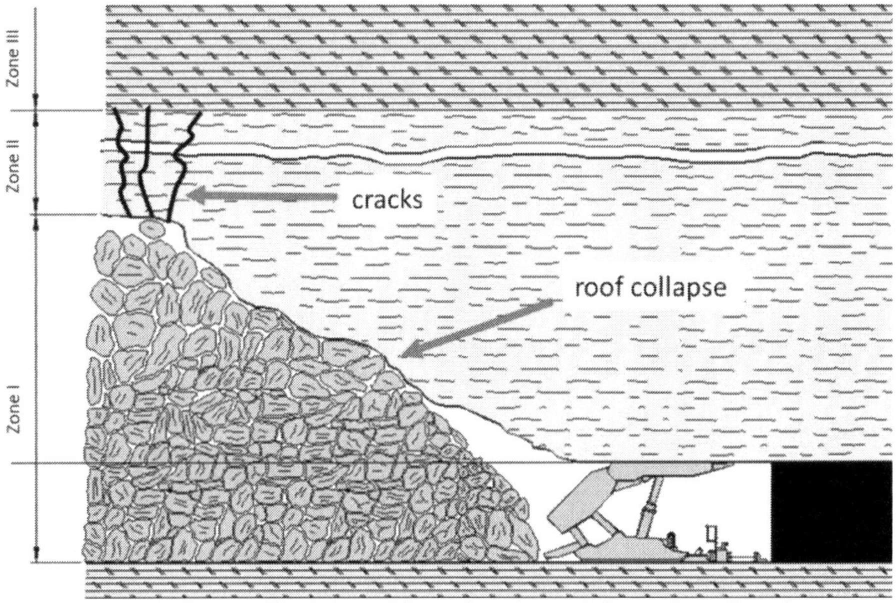

Figure 7. Longwall mining with roof caving and formation of cracks.

Another important consideration is the fact that compressed air in contact with coal may cause spontaneous combustion of coal and as a result uncontrollable fire. In order to reduce this risk remaining coal has to be separated from the storage site. If the coal mine is flooded or improperly sealed, coal mine flooding or uncontrollable inflow of water may reduce available volumes and increase moisture content of air which is not desired.

Even in regions where intensive mine operation has been carried at various depths (such as Upper Silesian Coal Basin in Poland) there is always a safety zone near the shaft which prevents it from skewing. In this zone roof support is designed for long term (permanent) use. Near shafts there are also large chambers for machine service or electrical equipment which could also enhance the volume of available workings (volume of single chamber itself is between 2000 and 7000 m^3). Other workings which could be used for compressed air storage include galleries or shafts and winzes. In Polish coal mines the typical roof support is the yielding steel arch support where cross-section area varies greatly between 5 up to over 36 m^2. The most common types have a cross-section area of 14.8 to 17.6 m^2. In Figure 8, a schematic view of typical yielding arch steel support crossing is presented.

IOP Conf. Series: Materials Science and Engineering **545** (2019) 012014 doi:10.1088/1757-899X/545/1/012014

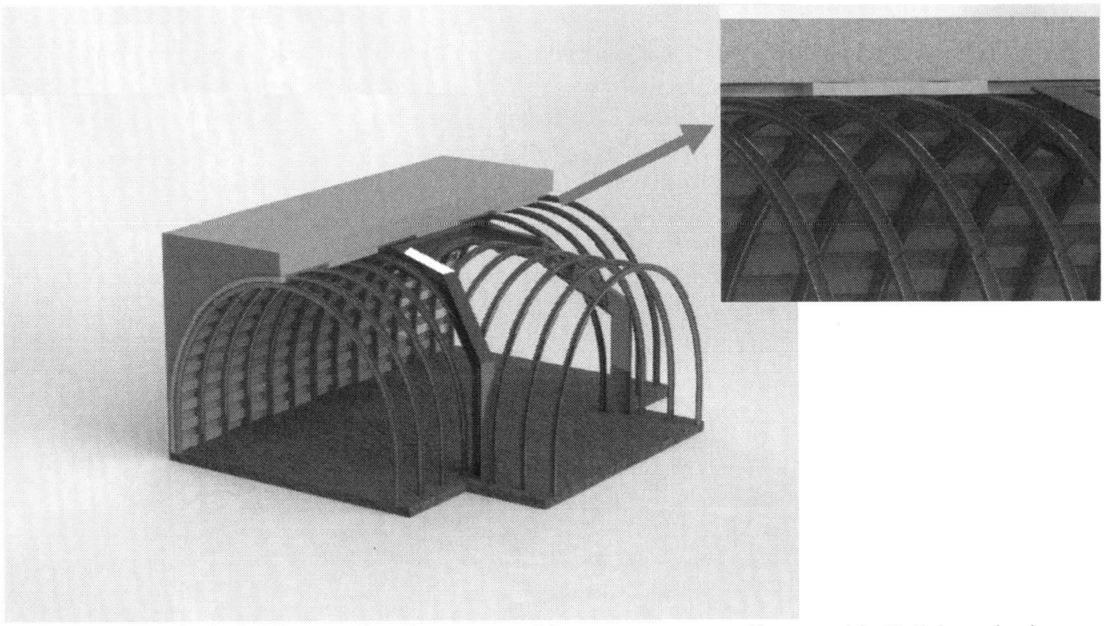

Figure 8. Typical yielding steel arch support with precast concrete liner used in Polish coal mines.

As it can be seen, the support lacks proper sealing since precast concrete is used as a liner behind the support. In order to convert such working into an impermeable air storage additional liner should be applied on the internal surface of the support (see Figure 9). The following materials could be used:

- shotcrete liners,
- geopolymer liners,
- polyurethane urea elastomers (PUU).

Before application of the liner a thorough check of support for corrosion, convergence control and integrity should be performed.

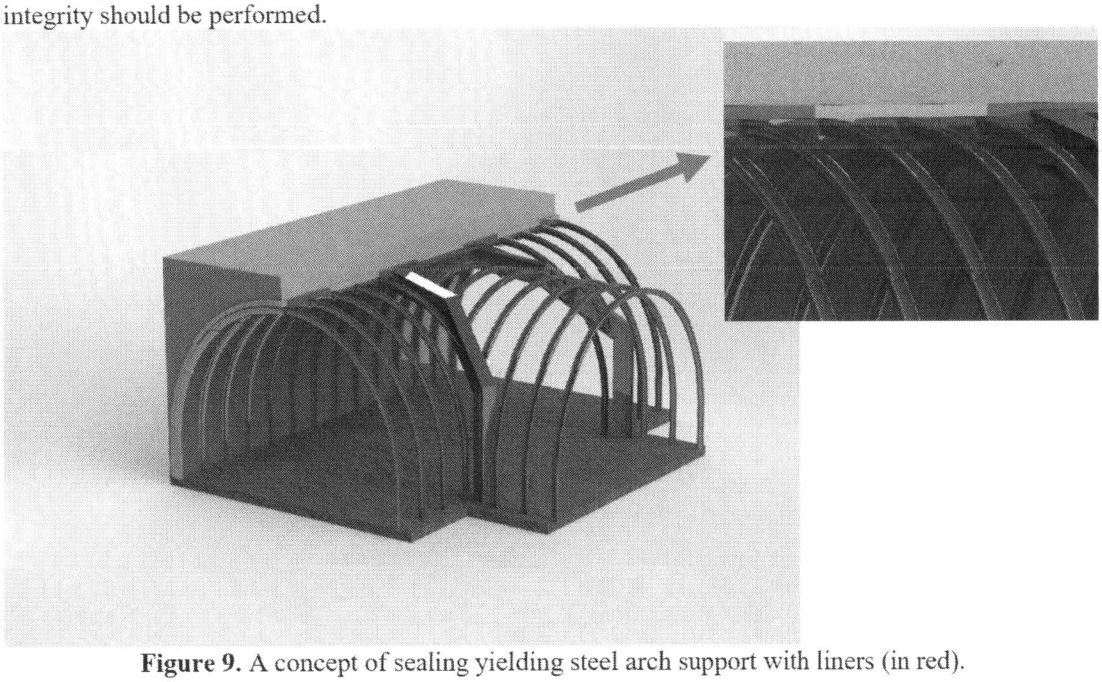

Figure 9. A concept of sealing yielding steel arch support with liners (in red).

IOP Conf. Series: Materials Science and Engineering **545** (2019) 012014 doi:10.1088/1757-899X/545/1/012014

6. Conclusions

With the increase in demand for energy storage technologies there is a need for seeking new solutions that could provide large scale energy storage. Compressed Air Energy Storage is one of the solutions, yet it requires leak-proof underground structures that will withstand high pressure and are located closely to the power sources. As one of the options, underground coal mines in European coal basins could be considered. Convenient location, well known geology of the site and already existing excavations are the key benefits of this idea. Nevertheless, there are many considerations that have to be taken into account and this is particularly leak-tightness and geomechanical stability of the overburden. Typical yielding steel arch support which is encountered in Polish coal mining could be used when a proper liner is applied. This concept however needs a thorough interdisciplinary studies that should integrate both energy efficiency issues as well as geological and geomechanical knowledge.

7. References

[1] European Commision 2019 *Renewable Energy Progress Report* (Brussels)

[2] Gielen D, Boshell F, Saygin D, Bazilian M D, Wagner N and Gorini R 2019 The role of renewable energy in the global energy transformation *Energy Strategy Rev.* **24** 38–50

[3] Anon 2015 Study on commercialisation of energy storage in European market to 2030 and beyond *Fuel Cells Bull.* **2015** 14–6

[4] Luo X, Wang J, Dooner M and Clarke J 2015 Overview of current development in electrical energy storage technologies and the application potential in power system operation *Appl. Energy* **137** 511–36

[5] Luo X and Wang J 2013 Overview of current development on compressed air energy storage *Sch. Eng. Univ. Warwick Coventry UK*

[6] Roland Marquardt 2014 Energy Storage – the Future of Renewables?

[7] Anon C.A.E.S. Technology/Hydrodynamics Group, LLC

[8] Milewski J, Badyda K and Szab\lowski \Lukasz 2016 Compressed air energy storage systems *J. Power Technol.* **96** 245–260

[9] Gillhaus A and Horvath P-L 2007 *Compilation of geological and geotechnical data of worldwide domal salt deposits and domal salt cavern fields* (Clarks Summit (PA): Solution Mining Research Institute)

[10] Lutyński M 2017 An overview of potential benefits and limitations of Compressed Air Energy Storage in abandoned coal mines IOP Conference Series: Materials Science and Engineering vol 268

[11] Menéndez J, Ordóñez A, Álvarez R and Loredo J 2019 Energy from closed mines: Underground energy storage and geothermal applications *Renew. Sustain. Energy Rev.* **108** 498–512

[12] Piessens K and Dusar M 2004 Feasibility of CO2 sequestration in abandoned coal mines in Belgium *Geol. Belg.* **7** 165–80

IOP Conf. Series: Materials Science and Engineering **545** (2019) 012014 doi:10.1088/1757-899X/545/1/012014

[13] K, Schultz 1998 *Gas Storage at the Abandoned Leyden Coal Mine near Denver, Colorado* (Raven Ridge Resources Inc)

[14] Cisek W, Dybciak A and Landsberg W 2001 Przebieg i doświadczenia z likwidacji kopalni“ Nowa Ruda” w aspekcie przekszta\lcenia wyroobisk do\lowych na potrzeby podziemnego magazynu gazu. *Przegląd Gór.* **57** 3–12

IOP Conf. Series: Materials Science and Engineering **545** (2019) 012015 doi:10.1088/1757-899X/545/1/012015

Magnetic-and-powder Method in Diagnostics of Welded Joints in Powered Roof Supports

S Szweda[1], J Mikuła[2] and S Mikuła[3]

[1]KOMAG Institute of Mining Technology, 37 Pszczyńska, 44-101 Gliwice, Poland
[2]Silesian University of Technology, Faculty of Mechanical Engineering, 18A Konarskiego, 44-100 Gliwice, Poland
[3]Silesian University of Technology, Faculty of Mining and Geology, 2 Akademicka, 44-100 Gliwice, Poland

sszweda@komag.eu

Abstract. A possibility of diagnosing the welded joints in the components of powered roof supports used in longwall panels of hard coal mines is analysed. It was found that the magnetic-and-powder method is most useful for that purpose. Devices which facilitate assessment of technical condition of welded joints in canopies and bases are discussed. Design and principles of operation of welded joints liquid detector adapted for operation in industrial conditions are presented. Two ideas of a magnetic inductor for technical diagnosis of powered roof support components by the magnetic-and-powder method is discussed.

1. Introduction

Conditions of the power roof support operation in longwall panels of hard coal mines cause their exposure to intensive destructive interactions, leading to the loss of functional properties and reduction in their technical safety. High load from the rock mass pressure, corrosion and abrasive wear are synergetic, mutually increasing the intensity of destructive processes. This applies in particular to canopies and bases, and especially to their welded joints [1, 2, 3]. The necessity of taking into account the synergy of destructive processes in assessment of the roof support virtual prototype is also emphasised by other researchers [4].

Variability of operational load to the roof support has a typical zero-pulsating character with a high amplitude of force and the main frequency resulting from the roof support operating cycle. These loads generate high strain, leading to formation and progress of fatigue cracks, typical for low-cycle load. Low cycle fatigue is characterized by the occurrence of plastic deformations in the crack zone and high rate of crack progress [5, 6].

The critical crack length at which the rapid acceleration of the crack progress begins, leading to a full fracture, in the case of low-cycle fatigue of steel components is much smaller than the critical crack length in the case of fatigue destruction of the components operating in high cycle ranges. The above applies in particular to the welded joints, which, by their nature, constitute strong constructional and structural notches causing the stress accumulation [7]. In addition, there are high welding stresses in the weld zone, often of unfamiliar distribution and character. In the case of spatially complex components such as canopies and bases, it is very difficult to remove welding stress using known technological procedures [5, 8].

In the described situation for maintaining the technical safety of the roof support, it is necessary to diagnose the technical condition of its components frequently, with special attention to the sensitive welded joints. This diagnosis often takes place in difficult conditions of underground mines. Therefore, the diagnostic methods used in these conditions must be simple and of low labour consumption while maintaining high efficiency, as well as low sensitivity to external operational disturbances. Among many defectoscopic methods of diagnosing the machine components, widely described in [9, 10], the penetrating and the magnetic methods deserve special attention.

Use of penetrants consists in covering the tested zones with special liquid that penetrates fatigue fractures and other surface defects. After a removal of excessing liquid and lighting the tested zone with special ultraviolet lamps, the penetrants remaining in fractures, are clearly visible, revealing the places of defects and cracks. Penetrative methods enable only a detection of defects and cracks occurring on the machine component surface and only in accessible places. This is the main disadvantage of these methods. An additional disadvantage of the methods is a need for a special preparation of the tested surface, which is difficult in the case of intensive corrosion and contamination of the component's surface in mine workings. The penetrants are usually prepared in an aerosol form, where the carriers are combustible solvents, what makes their use in the workings threatened by explosion hazard impossible.

The magnetic methods are free from the described disadvanteges. In particular, magnetic-powder methods are very useful [10]. They use magnetic powder (e.g. magnetite powder or other very fine ferromagnetic particles), which gathers with greater intensity in fatigue fractures of steel components, revealing the cracks or structural defects. The progressing fatigue cracks in the steel components, activate magnetically the fracture zones, so in many cases there is no need to use the external magnetic field. When the external magnetization is induced in the cracks zone or structural defects zone, the described effect of gathering the magnetic powder occurs more intensively, because the cracks and material defects change the lines of magnetic field. The main advantage of the magnetic-powder methods, in addition to their simplicity, is their high efficiency, even in the case of invisible cracks and subsurface defects.

2. Liquid detector dedicated to the magnetic-powder method

In magneto-powder defectoscopy, dry and the wet methods are distinguished [10]. With the dry method, a magnetic powder is applied to the tested surface after properly oriented magnetisation of steel component. A high consumption of magnetic powder and a need to prepare the surface by removing contaminants such as varnishes and layers of corrosion, which may be difficult in mine conditions, are the method disadvantages.

In the wet variant, the magnetic-powder method usually consists in applying a thin layer of quick-drying white base coat on the tested component, in a form of aerosol. Then, from the second container, a suspension of black finely dispersed magnetic powder in the liquid carrier is sprayed on the coating layer. However, the use of aerosol pressurised containers with flammable carrier substances cannot be accepted in the workings threatened by methane or coal dust explosion hazard.

In the situation described above, it may be better to use a special liquid detector whose design is shown in figure 1. The detector is a vessel (1) made of thin, flexible polymer (for example rubber) containing a transparent liquid carrier (2) with a small amount of finely dispersed magnetic powder.

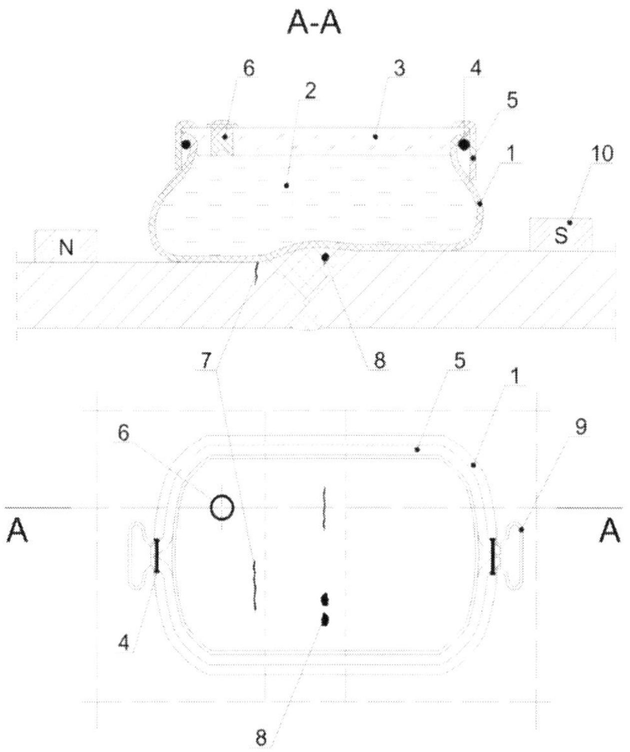

Figure 1. Liquid detector for localization of cracks and structural defects during testing the steel components by the magnetic-powder method (explanation of markings in the text).

The vessel (1) is closed by a transparent lid (3) with a seal (4) and a band (5). The elastomer plug (6) is used to refill the liquid carrier and vent the container inside. The band (5) is connected to the handles (9) for moving and fixing the detector.

Use of the detector requires shaking the container to obtain a homogeneous suspension of the magnetic powder in the liquid carrier. The detector is then placed on the diagnosed zone (e.g. weld). As a result of free sedimentation of the magnetic powder, it falls down to the vessel bottom, gathering with greater intensity within the zones of magnetic field disturbance generated by permanent magnets (10) or special inductors, a sample design of which is described in Chapter 3.

An example of using a liquid detector to assess the technical condition of a butt weld is presented in figure 1. Weld defects, e.g. fatigue cracks (7) in a weld or in a heat affected zone, slagging (8), gas bubbles, short-run castings - are signalled by aggregations of black magnetic powder on the container's white bottom. In the absence of weld defects, the bottom of the container has a uniform grey colour.

The hydrostatic pressure of the liquid carrier enables a good adhesion of the flexible walls of the container to the diagnosed surface. The surface of the tested component does not require any preparation, except for the possible removal of coarse impurities. Surface defects and cracks are best visualized when the magnetic field lines are perpendicular to the extent of defects or cracks. Therefore, it is advisable to diagnose the weld condition at least with two mutually perpendicular magnetization directions.

After a short time and after shaking it, the detector can be used again to test another weld. An observation of the vessel bottom (1) can be facilitated by the use of an optical magnifying glass in place of the transparent lid (3) and additional battery lighting. The detector facilitates a photographic recording of the weld damage condition for keeping the archieves of the damage or, for example, tracking the progressing fatigue crack. The described liquid detector enables a good assessment of the crack size or structural defect of the weld. To some extent, it is also possible to use foe an estimation of the depth of the subsurface defects.

3. Magnetic inductors adapted for assessments of welds in powered roof supports by the magnetic-powder method

Fully reliable results of diagnosing the technical condition of welds, especially for monitoring the development of fatigue cracks during the use of the powered roof support, are conditioned by the retention of repetitive magnetization parameters in subsequent weld condition evaluation cycles.

It is difficult to obtain the required repeatability of the magnetisation state of the tested roof support zones using the standard inductors with rigid permanent magnets or electromagnets with rigid pole pieces because the surfaces of the roof support components, especially welds, are usually exposed to intensive corrosive interactions, plastic deformations and abrasive wear. In addition, in the conditions of many mines, the powered roof support components are often covered by difficult-to-be-removed crust of mineral impurities. Such a condition of the roof support surface causes for example that the Solver inductors with four flat joints [9], or Pascal type DFN flaw detector [10], are of little use due to their inability for ensuring repeatable adhesion of their terminals to the tested surfaces. Besides, inductors with flat joints, at the connections of their segments, show a big loss of magnetic flux. These losses are also dependent on the angle of the joints bending.

Non-articulated inductor, presented in figure 2, is free from the disadvantages described above [11].

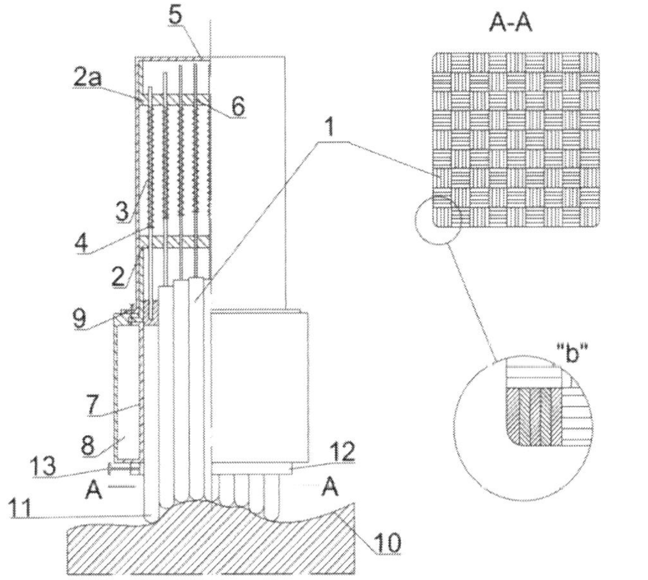

Figure 2. Non-articulated inductor for the components magnetization in magnetic-powder method [11] (expalnation of markings in the text).

The inductor core is made of a pack of loose rods (1) of the square or regular hexagon cross-section.

Each rod has a possibility of independent, free movement along its axis. The rods (1), made of transformer silicon steel, are loosely mounted in guides (2) and (2a) and held in starting position by individual springs (3), supporting one end with support plates (4), and the other end with internal guide bar (2a). The thrust plates (4) are clamped on the rods (1) guides (6). The cover (5) is permanently connected to the guide (2) and the coil bobbin (8) by screws (9).

Core rods can be monolithic or made of transformer sheets glued together (figure 2, fragment "b"). The side surfaces of rods (1) are covered by low-friction coating (e.g. Teflon) to reduce the friction force and to isolate from eddy current induced in rods after switching on the inductor electric power supply system.

Under pressure exerted on the tested surface (10) each rod slides partially to the inductor interior and the springs (3) secure a permanent contact with the tested surface. The current direction in the coil (8) should be selected in such a way that an electromagnetic force, acting on the core, is directed to the tested surface (10). After the test completion and removing the inductor from the tested object, all rods return to the starting position automatically. If necessary, there is a possibility to fix all the rods in a fixed or removable clamp (12) by the screw (13). That can be useful e.g. during testing the numerous identical components. The inductor can be used in combination with another the same inductor or with permanent magnets.

The method of using non-articulated inductors for testing welds is given in figure 3 [11]. A sample position of a crack or other flaw is marked by "x".

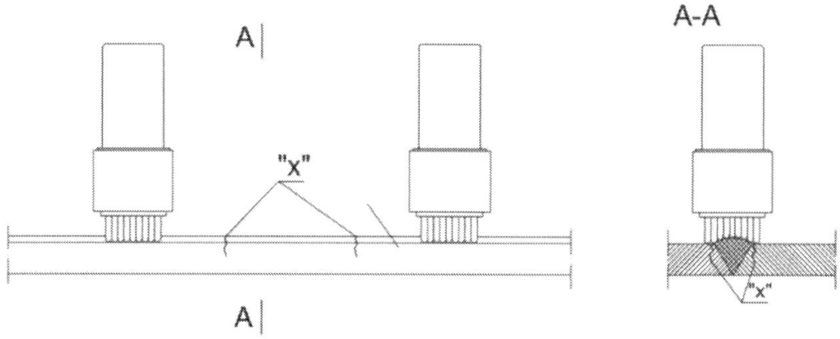

Figure 3. Example of using two non-articulated inductors for diagnosing the weld [11] technical condition. "x" – position of cracks.

The inductor ensures a permanent contact of all the rods with the tested surface securing a constant magnetization conditions despite a change in shape of the surface. Lack of articulated joints, eliminating a loss of a magnetic flux, enables supply the inductor with a current of lower voltage, what increases the work safety and reduces the power consumption. A non-articulated magnetic inductor (figure 2) can be used for a demagnetization of the tested object at the proper power supply, if necessary.

Another suggestion of an inductor design, intended for diagnosing the components, such as welds in powered roof support parts, is presented in figure 4. This inductor can be used in a version with permanent magnets (1a) or in a version with a coil (1b). An exchangeable and movable segment (2) is attached to one of the fixed parts of the inductor core (1), connected by the screw (12) with a pack permanent magnets (1a) or the coil (1b). Its position is fixed by a clamp (8) with a screw pressing (4). Strong tightening of the screw (4) enables good contact of the inductor's core components resulting in small magnetic flux losses. Pole pieces (5) are made of loose steel balls (9) of small diameter, installed in a flexible enclosure (6), made of, for example, thin rubber. The pole pieces are attached to the ends of the core by clamps (8) with clamping screws (7).

A small amount of oil lubricating the balls (6) was added to the enclosure interior for an anticorrosion protection and for a better adaptation of pole pieces to the tested surface (10), (11).

In the case of using the coil (1b) an adaptation of pole pieces to the tested components is enhanced by electromagnetic forces generated after switching on the electric feeder. The components (1) and (2) of the core are made of transformer sheet packs connected with glue and additionally with rivets. Clamps, screws and rivets are made of paramagnetic materials such as brass.

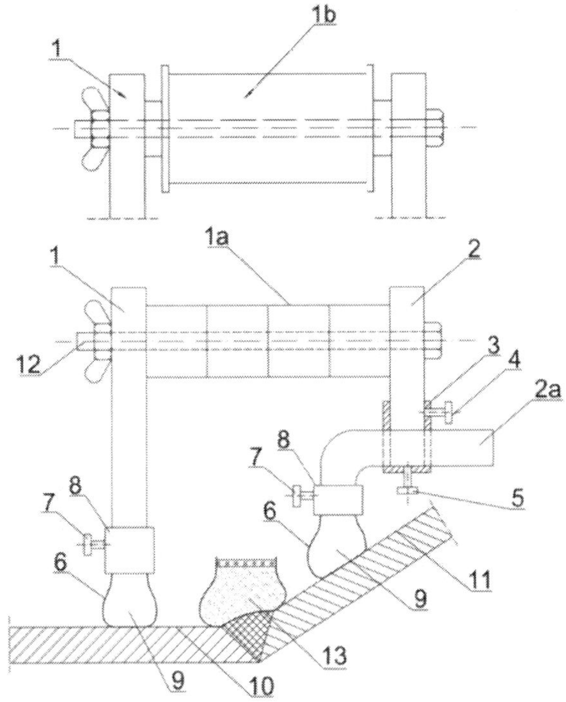

Figure 4. Inductor with a pack of permanent magnets (1a) or a coil (1b) for the components magnetization in the magnetic-powder method (explanation of markings in the text).

Examples of using the fluid detector (13), described in Section 2, for testing welds in powered roof support components of different spatial configuration and cooperating with the inductor with flexible pole pieces are presented in figure 4 and 5.

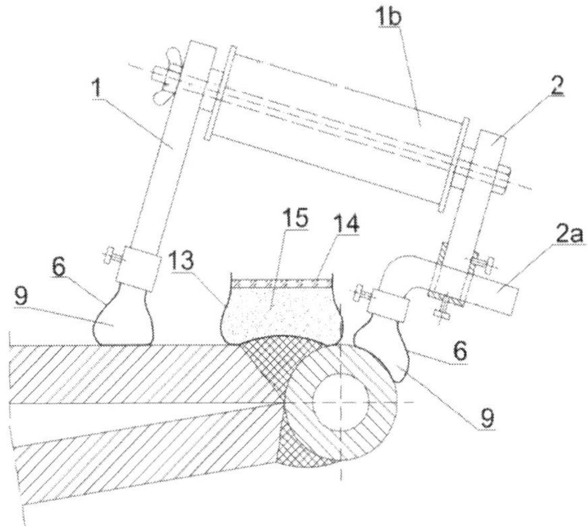

Figure 5. Example of using the inductor with flexible pole pieces and the liquid detector for diagnosing welds. (expalnation of markings in the text).

A new design of the inductor gives a good repeatability of a roof support magnetization during diagnosing of welds technical condition. Due to small losses in magnetic flux, the inductor ensures a high magnetization level at reduced voltage of power supply.

A liquid detector and magnetic inductors, described in this article, can be used for testing different components to detect cracks or structure flaws.

4. Conclusions

From a number of methods applied for technical diagnostics of powered roof support components, especially those realized in underground mine conditions, the magnetic-powder method in the wet variant is most useful.

The article describes a design of the fatigue cracks and structural defects liquid detector to be used for diagnosing the condition of powered roof supports components, with special emphasis on welded connections.

The described inductors, used for magnetization of roof support components, have many advantageous functions especially for diagnosing machine components in conditions of underground hard coal mines.

References

[1] Jaszczuk M and Markowicz J 2011 *Parametryczna metoda oceny stanu technicznego sekcji obudowy zmechanizowanej* ed. M Jaszczuk and J Markowicz (Gliwice: Wydawnictwo Politechniki Śląskiej)
[2] Szurgacz D 2017 Stress analysis of the powered roof support caving shield *Proc. XXV-th. Int. Conf. TEMAG Trwałość Elementów i Węzłów Konstrukcyjnych Maszyn Górniczych* Ustroń 19-21.10.2017 (Gliwice: Wydawnictwo Politechniki Śląskiej) pp 199-208
[3] Szweda S, Szyguła M and Mazurek K 2016 *Czynniki wpływające na postać konstrukcyjną i parametry techniczne sekcji ścianowej obudowy zmechanizowanej. Część 1* (Gliwice: Monografia Instytut Techniki Górniczej KOMAG)
[4] Winkler T and Tokarczyk J 2010 Multi-Criteria Assessment of Virtual Prototypes of Mining Machines. *Proc. of the World Congress on Engineering and Computer Science San Francisco USA* **2** pp 1149-53
[5] Gurney T R 1999 *Zmęczenie konstrukcji spawanych* (Warsaw WNT)
[6] Wyrzykowski J, Pleszakow E and Sieniawski J 1999 *Odkształcanie i pękanie metali* (Warsaw WNT)
[7] Robakowski T 1991 *Projektowanie złączy i zespołów spawanych* (Instytut Spawalnictwa Gliwice)
[8] Myśliwiec M 1972 *Cieplno- mechaniczne podstawy spawalnictwa* (Warsaw WNT)
[9] Kubiński W 2016 *Wybrane metody badań materiałów* (Warsaw PWN)
[10] Lewińska- Romicka A 2001 Badania nieniszczące – podstawy defektoskopii (Warsaw WNT)
[11] Szweda S 2017 Wzbudnik transformatorowy do badań magnetycznych elementów maszyn i urządzeń. Polska. Patent P 421490 z dn. 04.05.2017

IOP Conf. Series: Materials Science and Engineering **545** (2019) 012016 doi:10.1088/1757-899X/545/1/012016

Metals recovery from e-scrap using gravity, electrostatic and magnetic separations

D Franke and T Suponik

Silesian University of Technology, Faculty of Mining and Geology, Akademicka 2, 44-100 Gliwice, Poland

tomasz.suponik@polsl.pl

Abstract. The purpose of the work was to recover metals from e-scrap, in particular from mainboards of desktop computers, using gravity, electrostatic and magnetic separation. The article describes the method of comminution as well as the yield of size fraction after this process. In addition, the separation of the fraction of 0.32 - 0.56 mm with the use of two systems: I) electrostatic and magnetic separation, and II); gravity and magnetic separation, was presented. The feed to the magnetic separator for both systems was metals (high density products). The yield of fraction 0.32 - 0.56 mm obtained after comminution was 40.7% while the yields of the products obtained from the first system were as follows: in the case of electrostatic separation, 24.9% metals and 75.1% plastics and glass; and for magnetic separation 0.7% of ferromagnetics, 19.0% of diamagnetics and 5.3% of paramagnetics. The yields of the products in the second system were: in the case of gravity separation, fraction with a density $\rho > 2.96$ g/cc = 33.0% and fraction with a density $\rho < 2.96$ g/cc = 67.0%; and for magnetic separation 0.8% of ferromagnetics, 25.3% of diamagnetics and 6.9% of paramagnetics.

1. Introduction

For the production of electric and electronic equipment that are used in many areas of our lives, large amounts of valuable metals are needed. After using them for several years, these products become waste. Due to the financial value of metals and in accordance with the circular economy, many researchers have undertaken to recover theses metals. The presence of all metals in waste is 60.6%, of which 47.9% are iron and steel, 7.0% copper, 4.7 % aluminum, 1.0 non-ferrous [1].

Printed circuit boards (PCBs) of PC are particularly interesting due to the content of metals in them [2, 3], as well as their large amount in all waste of electric-electronic equipment. PCB can contain up to 20% Cu and 250 g/ton Au, which are very high i.e. 25-250 - fold for gold and 20-40 - fold for copper when compared with gold ores and copper ores, respectively [4]. A motherboard is the main printed circuit board found in general purpose computers and other expandable systems. They contain the following metals: copper 11-14%, iron 4.5-8%, tin 2-4%, aluminum 1.5-3%, lead 1.5-2%, nickel 1-2.5% and zinc 0.5-1%. The precious metals present in the motherboard are mainly: 0.005% gold, 0.008% silver, 0.002% palladium, 0.004% platinum and negligible content of some rare earth elements. Other materials are mainly plastics, such as polymers, polyesters, phenol formaldehyde, halogenated polymers, and nylon or polyurethane [5, 6]. It should be noted that the waste processing also has an environmental dimension in addition to financial considerations.

The methods and devices of gravity, magnetic and electrostatic separation have been presented in many works [7-12].

To summarize, technological progress and the need for more and more powerful computing units make it necessary to constantly exchange computers for new ones. This creates a large amount of electronic scrap, which contains much larger amounts of metals than in natural deposits. At the same time, they contain metals harmful to the environment. Therefore, in the work the gravity, electrostatic and magnetic separation was used to recover metals from motherboards originating from desktop computers. For this purpose, four motherboards from different manufacturers were applied.

2. Research methods

In order to recover metals from mainboards, the methodology presented in figure 1 was applied. Prior to separation, preliminary work was carried out consisting in preparing motherboard for two subsequent systems. The purpose of these systems was to separate metals from plastics and glass. Electrostatic and magnetic methods were used in the first system, while gravity and magnetic in the second (see figure 1). The preliminary work consisted of material fragmentation and their division into grains of the same size.

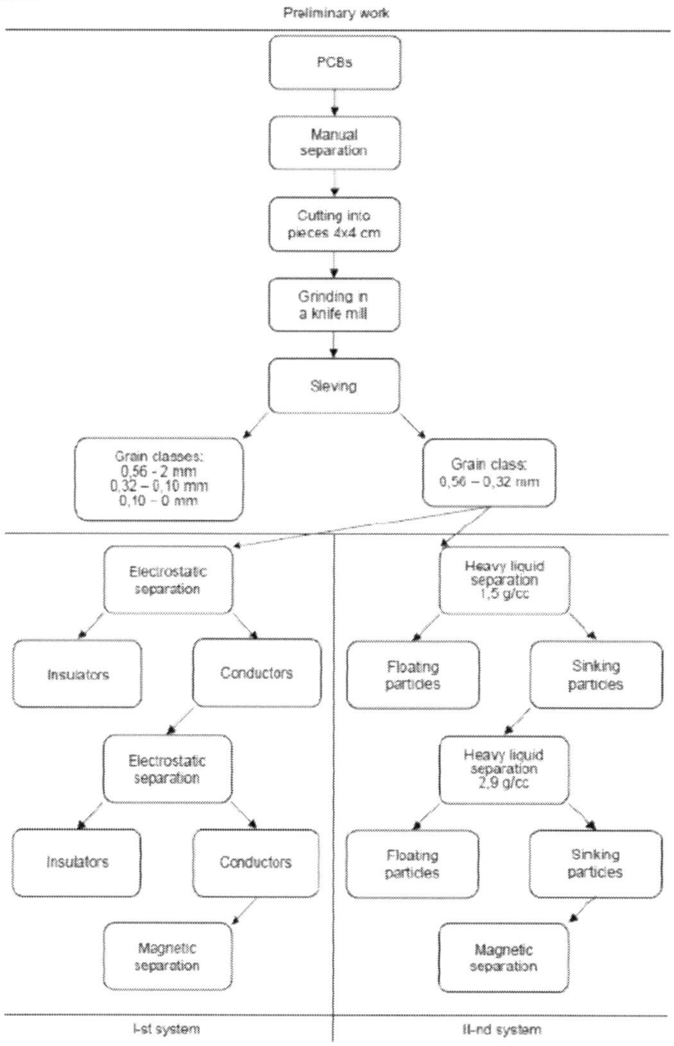

Figure 1. Research methodology.

2.1. Preparation for grinding

Printed circuits boards have many components that can cause problems when grinding or do not contain metals and at the same time can be easily removed with simple methods. Therefore, the first step was manual separation of selected parts of PCBs using basic tools such as screwdrivers and pliers. The following parts of PCB was removed: batteries, capacitor, inductors, quartz crystals, relays, resistors, transistor, integrated circuits and others (screws, connectors, switches and etc.). Then, the pre-cleaned PCBs were cut into 4x4 cm pieces due to the limitations of the knife mill.

2.2. Grinding

PCBs consist of many materials such as flexible plastic, hard glass and ceramics and various types of metals. In addition, each element has a different shape and thickness. Rotational speed, type of crushing and degree of crushing are important parameters when choosing the right type of grinding device. In the research, a knife mill was used to grind materials with hardness below 4 on the Mosh scale. Table 1 presents the parameters of this grinder. The mill is equipped with 3 sets of knives, 3 knives placed on the shaft and 3 knifes on the grinding chamber.

Table 1. Parameters of the knife mill [13].

Power	2.2 (kW)
Rotation speed	2815 (rpm)
Preliminary grain	<40 (mm)
Final grain	Depending on used sieve
Standard sieve diameter	3; 2; 1; 0.425 (mm)
Efficiently	Depending on sieve approx. 10 (kg h^{-1})

Grinding was carried out at a rotation speed of 2815 rpm using hardened steel blades. In order to obtain the particle size suitable for the application of magnetic and electrostatic separation, a perforated sieve with a mesh size of 2 mm was used.

2.3. Screening

To prepare material for separations, four types of screens with different mesh sizes: 2.0 mm; 0.56 mm; 0.32 mm and 0.10 mm were used. The particle size 0.56 - 0.32 mm was used for the tests presented in this paper. The processing of other particles will be analyzed in subsequent studies.

2.4. Electrostatic separation

Electrostatic separation was carried out using the Boxmag-rapid Ltc separator. Electrostatic separation uses the difference in the surface conductivity of different materials. In the Boxmag separator, the material is fed to the belt of rotating shaft where it is subjected to corona discharge using a high voltage flowing through the electrode wire. This causes ionization of the air, thanks to which it is possible to charge material particles on the shaft. Particles with high surface conductivity quickly lose charge and are ejected from the rotating shaft by centrifugal force. The remaining particles that have not lost their charge are removed by a brush at the rear of the shaft. The main parameters of separation are: rotational speed of the shaft, voltage flowing through the electrode and its distance from the shaft, and the size of the material and feeding speed [14].

As a result of research conducted on the analyzed material, it was found that the most optimal parameters of separation in terms of efficiency and purity are: voltage flowing through the electrode = 20 kV, speed of shaft rotation 50 rpm and distance of electrode from the shaft of 5 cm.

2.5. Gravity separation

In subsequent works, it is planned to verify the efficiency of separation using the difference in grain density, e.g. with the use of concentrating table or cyclofluidic separator. Due to too small sample of

IOP Conf. Series: Materials Science and Engineering **545** (2019) 012016 doi:10.1088/1757-899X/545/1/012016

the analyzed material, only a simplified density analysis was carried out. Two liquids with densities of 1.5 g/cc ($ZnCl_2$ solution) and 2.96 g/cc (1.1.2.2-tetrabrom-eathan $Br_2CHCHBr_2$) were used.

2.6. Magnetic separation

The separation was made using the magnetic susceptibility of the material and the magnitude of the magnetic force. For this purpose, a disk magnetic separator was used, which ensures heterogeneity of the magnetic field. The material placed on the moving belt is affected by the ever-larger magnetic field. Grains of material with a higher magnetic susceptibility are lifted from the belt and placed in the receiver's containers by means of a chutes. The separator allowed to separate the material by means of two different magnetic forces produced by electromagnets [15]. Grains with the lowest magnetic susceptibility pass under magnetic plates and fall from the conveyor belt at the end of the separator. Additionally, at the beginning of the separator there is a permanent magnet that captures the ferromagnetic material.

The density of separated products in all tests was specified using pycnometer method in line with the PN-88/B-04481 standard. The content of ferromagnetic materials in the samples was carried out using the Satmagan 135 meter by Rapiscan Systems.

3. Results and discussion

3.1. Manual separation and grinding

Four motherboards from different manufacturers and with different masses were used in the tests. On average, about 45% of the mass of the motherboards were waste that had been manually removed (see table 2). This is a large amount of waste that also contains metals. Therefore, the recovery of metals from these wastes will be considered in subsequent tests.

Table 2. Results of manual separation of motherboards.

	Mass of the motherboards, (g)	Mass of the motherboards after manual separation, (g)	Percentage of waste, (%)
	632.4	355.0	44
	558.6	310.0	45
	497.2	269.6	46
	496.8	265.6	47
Total	2185.0	1200.2	45

During the grinding of elements with a larger content of plastics that could not be removed manually, an increase in the temperature in the material and the grinding chamber was noticed. Therefore, it is recommended to use liquid nitrogen for grinding more material.

3.2. Sieving

Table 3 shows the results of the granulometric composition after grinding. The lack of grains > 2 mm indicates good quality of grinding. During this process, in a knife mill a sieve of 2 mm was used. Two grain classes represent almost 80% of the sample. They are 0.32 - 0.56 mm and 0.10 - 0.32. The higher one was undergone to separation in the system I and II (see figure 1).

Table 3. The results of the screening tests of comminuted motherboards.

Grain class, mm	Class weight, (in % by mass)
> 2	0.0
0.56 – 2.00	7.9
0.32 – 0.56	40.7
0.10 – 0.32	38.4
< 0.10	13.0

3.3. Electrostatic separation

Electrostatic separation was carried out only for system I (see figure 1). Due to the difficulties associated with the separation of materials caused by the permanent connection of plastics and metals, a two-stage separation was carried out. In the first step, grains consisting of plastics and glass were separated from grains consisting of metals and their combinations with plastics. Then, the electrostatic separation of the metal product was again performed under the same separation conditions that allowed to separate metals from the grains in which they were combined with plastics. As a result, 25% of metals were obtained with a small amount of impurities, as evidenced by the density of this product at 11.17 g/cc (table 4).

Table 4. Results of electrostatic separation.

Electrostatic separation		Weight, (in % by mass)	Density, (g/cc)
1st	Plastics	69.7	3.86
	Metals	30.3	10.50
2nd	Plastics	5.3	7.43
	Metals	24.9	11.17

3.4. Heavy liquid separation

Density separation was performed only for system II (see figure 1). For technical reasons, it was conducted into two stages. In the first stage grain separation in $ZnCl_2$ solution (density 1.5 g/cc) was performed, while in the second in tetrabrom-eathan (density 2.96 g/cc). A high density value for the fraction of <1.5 g/cc indicates the difficulty of separation for fine grains resulting from the high solution viscosity. Grains with a density > 2.96 g/cc made up ca 1/3 of the total material, and their density was almost 10 g/cc.

Table 5. Density distribution results.

Separation in heavy liquid		Weight, (in % by mass)	Density, (g/cc)
1st	Fraction < 1.50 g/cc	56.9	3.16
	Fraction > 1.50 g/cc	43.1	7.25
2nd	Fraction < 2.96 g/cc	10.1	-
	Fraction > 2.96 g/cc	33.0	9.63

3.5. Magnetic separation

Magnetic separation was the last stage of research conducted in both systems. The separator parameters were the same. In the case of I system , the paramagnetic mass was sufficient to assess the density and content of ferromagnetic particles in this product, whereas in the system II the mass was

IOP Conf. Series: Materials Science and Engineering **545** (2019) 012016 doi:10.1088/1757-899X/545/1/012016

too low. Therefore, in the system II, paramagnetic products with higher and lower magnetic susceptibility have been combined.

The analysis of the results presented in table 6 shows that the I system was more efficient in separating metals from plastics, because the density of all products was higher by about 1.5 g / cm3 than in the system II. The densities of precious metals were presented among others the paper by Tuncuk et al. [4]. Analysis of the content of ferromagnetic materials confirmed the correctness of magnetic separation in both systems. F

Figure 2 shows the products after the separation processes. The shape and color of the grains indicate that the separation of plastic from metals was done precisely. Diamagnetics are also significantly different from paramagnetics, which proves the effectiveness of separation.

Table 6. Results of magnetic separation (for both systems)

Products susceptibility to magnetic fields		System I			System II		
		Weight, (in % by mass)	Density, (g/cc)	Content of ferromagnetic, (%)	Weight, (in % by mass)	Density, (g/cc)	Content of ferromagnetic, (%)
ferromagnetic		0.7	11.33	90.5	0.8	10.63	-
paramagnetic	lows magnetic force	4.0	13.38	1.0	4.8	11.61	0.2
	higher magnetic force	1.3	13.22	0.0	2.0		
diamagnetic		18.7	10.58	0.0	25.3	9.08	0.0

In terms of value and suitability, each product is important. Diamagnetics can contain such metals as: gold, silver, copper or lead, while paramagnets palladium, tin, chromium, aluminum, as well as rare earth metals. In turn, ferromagnetics can contain iron, cobalt and nickel.

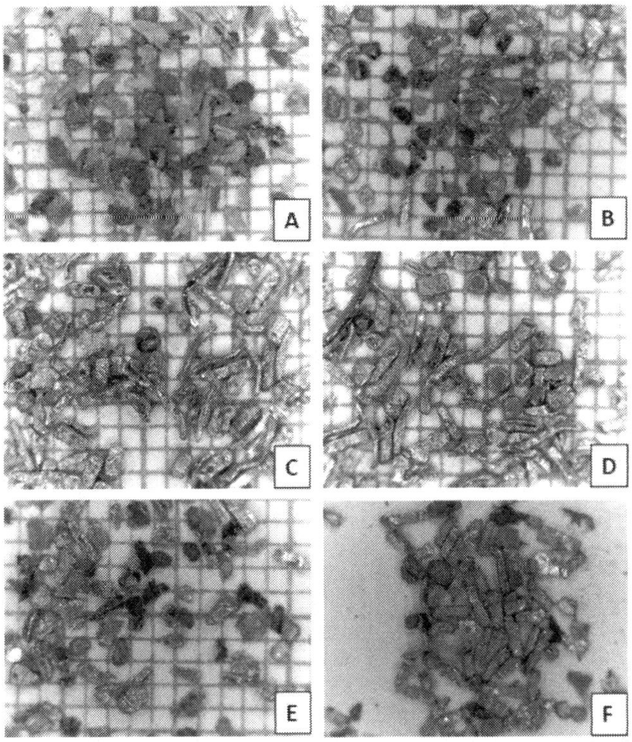

Figure 2. The grains of: A - plastics after the first electrostatic separation, B - metals after the second electrostatic separation, C - paramagnetic (low magnetic force), D - paramagnetic (high magnetic force), E – diamagnetic, F – ferromagnetic.

4. Conclusions
The following conclusions can be drawn on the basis of the research:
- the yield of fraction 0.32 - 0.56 mm obtained after grinding a motherboards in the knife mill was 40.7%,
- the separation of metals from plastics in the fraction of 0.32 - 0.56 mm is effective in both analyzed processing systems, i.e. in electrostatic and magnetic separations, as well as in gravity and magnetic separation,
- the yields of the products obtained from electrostatic and magnetic separation were as follows: in the case of electrostatic separation, 24.9% metals and 75.1% plastics and glass; and for magnetic separation 0.7% of ferromagnetics, 19.0% of diamagnetics and 5.3% of paramagnetics,
- products outputs in gravity and magnetic separation were as follows: in the case of gravity separation, the fraction with a density $\rho > 2.96$ g/cc amounted to 33.0% while the fraction with $\rho < 2.96$ g/cc 67.0%; and for magnetic separation 0.8% of ferromagnetics, 25.3% of diamagnetics and 6.9% of paramagnetics.

References
[1] Widmer R, Oswald-Krapf H, Sinha-Khetriwal D, Schnellmann M and Böni H 2005 Global perspectives on e-waste *Environmental impact assessment review* **25**(5) pp 436-458
[2] Hagelüken C 2006 Recycling of electronic scrap at Umicore precious metals refining *Acta Metallurgica Slovaca* **12** pp 111-120

[3] Yazıcı E, Deveci H, Alp I, Akcil A and Yazıcı R 2010 Characterisation of computer printed circuit boards for hazardous properties and beneficiation studies
[4] Tuncuk A, Stazi V, Akcil A, Yazici EY and Deveci H 2012 Aqueous metal recovery techniques from e-scrap: hydrometallurgy in recycling *Minerals engineering* **25**(1) pp 28-37
[5] Oh CJ, Lee SO, Yang HS, Ha TJ and Kim MJ 2003 Selective Leaching of Valuable Metals from Waste Printed Circuit Boards *Journal of the Air & Waste Management Association* **53**(7) pp 897-902
[6] Menad N and Van Houwelingen J 2011 Identification and recovery of rare metals in electric and electronic scrap: a review *Tirtheenth International Waste Management and Landfill Symposium*pp 10 p
[7] Drzymała J 2007 *Mineral processing* Ofic Wyd PWr
[8] Svoboda J 1987 *Magnetic methods for the treatment of minrals* Amsterdam Elsevier
[9] Kelly E G and Spottiswood D J 1982 *Introduction to mineral processing*
[10] Laskowski J and Łuszczkiewicz A 1989 *Przeróbka kopalin wzbogacanie surowców mineralnych* Dział Wydawnictw Politechniki Wrocławskiej
[11] Zhang S and Forssberg E 1998 Mechanical recycling of electronics strap-the current status and prospects *Waste Management & Research* **16**(2) pp 119-128
[12] Cui J and Forssberg E 2003 Mechanical recycling of waste electric and electronic equipment: a review *Journal of hazardous materials* **99**(3) pp 243-263
[13] www.testchem.pl
[14] Bedeković G, Salopek B and Sobota I 2011 Electrostatic separation of PET/PVC mixture *Technicki vjesnik* **18**(2) pp 261-266
[15] Blaschke S 1972 *Przeróbka mechaniczna kopalin* Wydawnictwo Śląsk

IOP Conf. Series: Materials Science and Engineering **545** (2019) 012017 doi:10.1088/1757-899X/545/1/012017

Utilization of methane deposits from the liquidated mines in the aspect of environmentally friendly energy technologies

S Trenczek[1] and E Krause[2]

[1]Professor, Institute of Mining Technology KOMAG, Gliwice, Poland

[2]Professor, Central Mining Institute, Katowice, Poland

strenczek@komag.eu

Abstract. At the beginning of the article main factors occurring in polluted atmospheric air and their connection with the methods of energy production are discussed. Various energy sources in the aspect of energy production technologies, including primary energy sources are presented. Two main actions taken to improve the quality of atmospheric air in Poland are described: Clean Air Programme, assuming a subsidy for an exchange of individually-used coal boilers for ecological ones or for gas heating, and the Polish Energy Policy until 2040, which assumes a gradual reduction in the share of hard coal and lignite in energy production. Methane as a mineral accompanying coal seams as well as ways of its recognition and management are presented. An example of the conducted research project shows a significance of the resource recognition and dynamics of the deposit in planning an economically justified exploitation of the deposit from closed down mines. Finally, there is hope that due to a further process of efficient restructuring of the Polish mining industry, the importance of using methane from closed down mines as a green energy source will increase.

1. Introduction

An environmental protection, including the fight against smog, gains more and more importance in the European Union. In Poland, a need to take such actions is confirmed by a control carried out by the Supreme Audit Office [1], which shows that the main cause of insufficient air quality in Poland is the emission of suspended dust and benzo(a)pyrene from domestic furnaces and local coal-fired boilers. The concentrations of benzo(a)pyrene, which should be at most according to WHO standards (World Health Organization) 1 ng/m^3 of air, are exceeded in Poland more than eight times. A slightly better situation - as reported by the Supreme Audit Office - is in the case of PM10 and PM2.5 air pollution, although their concentrations vary in different parts of our country. Therefore, as regards heating buildings and houses, the government of the Republic of Poland encourages a replacement of old boilers with more ecological ones, as well as home insulation under the Clean Air Programme [2], for which PLN 103 billion are allocated. Replacing furnaces and introducing thermo-modernization of houses are aimed at improving the energy efficiency of single-family buildings and reducing an emission of dust to the atmosphere.

It should be emphasized, however, that these are not the only actions aimed at reducing the so-called low emission, mainly due to the fact that environmental protection is not conducive to energy production from solid fuels. The energy system in Poland is based on hard coal and lignite. Therefore, the government of the Republic of Poland undertakes measures aimed at abandoning a production of

IOP Conf. Series: Materials Science and Engineering **545** (2019) 012017 doi:10.1088/1757-899X/545/1/012017

energy from these raw materials, through increasing a share of pro-ecological energy, including the energy produced from natural gas and nuclear energy [3].

If a possibility of using methane as the so-called clean fuel is considered, although the scale is relatively small - it is also worth using it even on such a small scale.

2. Plan of energy sources

A development of science and technology has caused that for a production of energy, people use many solutions, including the basic energy sources occurring on the Earth, such as the sun, water or wind, i.e. renewable sources [4]. Among them, technologies allowing a separate generation of electricity and of thermal energy, as well as technologies generating both electricity and thermal energy, are distinguished - Table 1.

Table 1. Type of produced energy in the context of energy production technology [4]

Type of produced energy	Technology of energy production	Primary source of energy
electric	water power plants	water
	wind power plants	wind
	power plants using ocean currents	sun - heating ocean waters, Earth, atmosphere
	photovoltaic cells and solar power plants	sun - solar radiation
	tidal power plants	gravity - tidal waters associated with the moon phase
thermal	heat pumps	sun - solar radiation
	collectors and thermal solar power plants	
electrical and heat	wind power plants	wind
	combined heat and power plants	sun - biomass
		Earth- geothermal sources

As it is well known, a production of electricity and heat in Poland is mainly carried out by large, institutional suppliers. According to the data of the Statistics Poland [5], in 2017 in Poland, 74.6 million tonnes of coal were consumed for the energy production, while for the energy sector (power plants, combined heat and heating power plants) 59.6% of consumption and 23.6% for industry and construction, including the consumption by mines and the consumption by coking plants. Significant consumers of coal included households (13.7%). In turn, a consumption of natural gas (excluding consumption for technological needs of the gas sector) amounted to 628.5 PJ in 2017.

For many years, the share of coal has been significant in the production of electricity in Poland. It amounted to 46.4% in the case of hard coal, and 30.6% in the case of lignite. Moreover, electricity was also generated from wind - 8.7%, gas fuels - 5.7%, biomass – 3.1% and other carriers - 5.2%. It can be noticed that there was a constant trend of changes in the proportions of the share of individual energy sources in the period 1999-2017, as illustrated in Figure 1 [6].

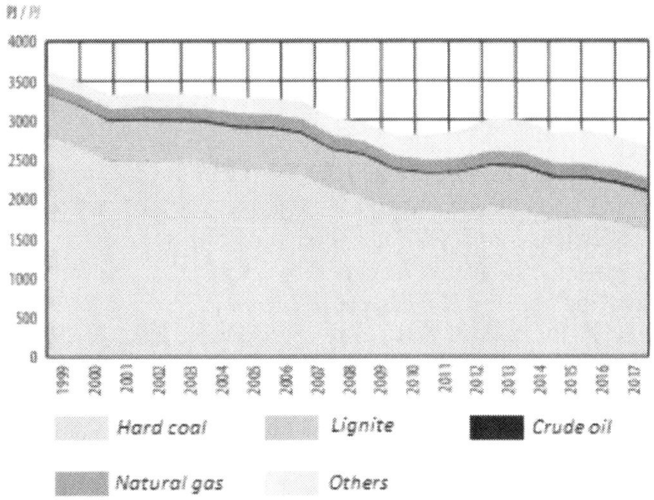

Figure 1. Primary energy acquisition [6].

3. Pro-ecological activities in the energy generation sector

As it has already been mentioned, some actions have been taken in Poland aimed at moving away from hard coal and lignite energy generation to increase the share of pro-ecological energy. The first but significant results are to be experienced until 2040, which is reflected in the document "Energy Policy of Poland until 2040" (PEP2040) [3]. It is a strategy setting directions for a development of the energy sector - including the tasks necessary for an implementation in the short-term - in full compliance with the European Union strategic documents.

The forecasting data included in PEP2040, are the result of optimization modelling of the shape of the power balance in the 2040 perspective, taking into account (among others) an increase in installed net power from about 40 GW to 72.6 GW in 2040 and an increase in electricity production from approx. 165 TWh to 232 TWh in 2040. PEP2040 assumes that it is impossible to deviate from fossil energy sources completely. However, it is assumed that there will be an optimization of the use of domestic energy resources, based on a rational use of energy resources, i.e. hard coal, lignite, natural gas, crude oil, biomass and non-agricultural waste, ensuring, inter alia, a profitability of the mining sector, a rational exploitation and innovations in an extraction and use of raw materials. The need to expand a generation and network infrastructure of electricity producers was also taken into account, aimed at covering the demand for electricity with generation capacities, other than conventional coal in a stable, flexible and ecological way.

Some important activities include:

- an implementation of nuclear energy, aimed primarily at reducing the emission level of the energy sector and system operational safety - the first nuclear unit with the capacity of 1-1.5 GW is expected to be operational by 2033, and the next five ones - by 2043 (in total approx. 6-9 GW),
- a development of renewable energy sources aimed at reducing the emission intensity of the energy sector and a diversification of energy generation – the share of 21% of RES is expected in the final gross energy consumption in 2030,
- a development of heating and cogeneration, aimed at ensuring universal access to heat and low-emission heat generation throughout the country, through system heating and individual heating.

Drawing attention by the PEP2040 creators to a development of cogeneration gives hope for increasing the scope of activities related to obtaining methane and its economic utilization.\

4. Methane as a source of energy

Currently in Poland, for the energy production - although on a small scale - methane is also used. It is a mineral accompanying coal seams. However, to use it, it must be captured earlier, which it can be done in several ways, depending on what kind of coal deposit and the geological and mining environment are involved.

Due to the method of capturing coal seam methane, [7] the following cases can be distinguished:

- methane from coal seams unaffected by mining exploitation, i.e. Coal Bed Methane (CBM), which is obtained before starting the operation of a given seam,
- methane coming from coal seams disturbed by exploitation, released to mining excavations during hard coal mining, i.e. Coal Mine Methane (CMM), which is partially captured by demethanization of the deposit, and partly released into the ventilation air (Ventilation Air Methane - VAM),
- methane from coal seams, released after their exploitation to mine workings, so-called Abandoned Mine Methane (AMM), located in the areas of liquidated mines.

In Poland, due to a very low carbon permeability of the Upper Silesian Coal Basin, methane is not included in the pre-operational phase (CBM). However, methane content is relatively high in the operational and after-operational phase (CMM). According to the data of the Industrial Development Agency [8], an increasing amount of separated methane (CMM) is captured by surface demethanization stations - in 2016, approx. 346.6 million m^3 were captured. Some of this amount of methane is consumed by mining enterprises - in 2016 it was 47.6 million m^3 - which enabled to produce over 270.3 thousand MWh of electricity and almost 760.3 thousand GJ of heat.

However, as regards the production of energy, statistical data do not include information about the use of methane from liquidated mines, probably because it is negligible in the total amount of energy production. The fact is that an interest in such methane acquisition increases - in 2012 there were two mining plants in Poland operating AMM, in 2017 - three, and an exploration and identification of methane deposits from coal seams were conducted in Poland on the basis of eight concessions granted by the Minister of the Environment. The gas, extracted from liquidated mines, is managed either directly by the methane extraction plant or after transferring it to the entrepreneur who deals with the generation of energy in cogeneration aggregates.

It should also be added that in contrast to the 1990s, when during the intensive restructuring of the Polish mining industry, there was no vision of methane use after a liquidation of mines, now this issue is already addressed at the mine decommissioning planning stage. There is a chance to develop system solutions for AMM gas recognition. It seems to be justified in every respect, as the number of mines planned for liquidation will constantly increase. This will be served by the mentioned entry in PEP2040, referring to methane, which will probably cause an interest in AMM gas to increase even more. This is also served by the amendment to the Act on the Functioning of Hard Coal Mining and the Act Amending the Geological and Mining Law and Certain Other Acts, introducing solutions to facilitate a development of methane obtained from hard coal deposits within a liquidated mining plant or its designated part [9]. This facilitates an extraction of methane from coal seams (as an accompanying mineral) of liquidated mines, which means that a mining enterprise, dealing with the liquidation of mines, does not have to have a hydrocarbon license for obtaining methane from liquidated mining plants or their designated parts. This will facilitate a development of methane exploitation, while at the same time ensuring the safety of neighbouring mines and the environment.

5. Economics of using methane from liquidated mines

Current experience shows that the best way to develop AMM is to generate energy in cogeneration aggregates, which are highly specialized devices for combined production of electricity and heat. Usually it is a combination of an internal combustion engine, a generator, a set of heat exchangers and a control system that allows the unit to be controlled locally, remotely via a computer or a mobile phone. Contrary to other alternative ways of producing electricity, the greatest benefit of using a cogeneration is that it is possible to produce as much energy as it is needed at a given moment. For this reason, cogeneration is sometimes referred to as a controlled source of energy.

However, energy production in CHP cogeneration aggregates must be economically justified, which is influenced by many factors. One of them is the time of potential efficient extraction of methane from a given deposit.

The energy production planning in cogeneration aggregates is based on a lot of data - regarding both the parameters of the cogeneration aggregates and the parameters characterizing the methane deposit. For example, if the extracted methane would be consumed in the ENGINE Caterpillar 3516 1.136 MW, its optimal performance parameters should be taken into account - combustion of 5 m^3/min gas at 100% CH_4 - and limit tolerable performance parameters - combustion of 15 m^3/min gas with a minimum concentration of 33% CH_4, as well as an economically optimal engine operation time - min. 8.400 h/year. These parameters set the right requirements for methane resources in the reservoir, while the exploitation resources, understood as 67% of industrial resources, are more important than industrial resources. Only such resources, due to limitations of the minimum methane concentration, can be used in an internal combustion engine in its continuous, safe operation [10].

6. Dynamics of the methane deposit as a determinant of the method of exploitation

It is important to adopt an appropriate business model for the effective exploitation of the AMM methane deposit when planning energy production in cogeneration aggregates. It is about the choice of the deposit extraction method: stable - long-term, realized with one aggregate, at specific fixed costs, or dynamic - short-term, implemented with the use of two or more aggregates, with other fixed costs. However, to conduct such considerations, the methane deposit should also be recognized in terms of its dynamics.

In the case of a seam with gas collecting features - and a vast majority of deposits in closed down mines have such features (due to an occurrence of unsubstituted, uninhabited mine workings) - exploited with one bore-hole, e.g. A bore-hole first methane deposit performance (W_{ZM}) [m^3/t_i] is determined from [11]:

$$W_{ZM} = (W_P \cdot p_m) \tag{1}$$

where: W_P - efficiency of the gas pumping station (amount of gas extracted from the bore-hole) [m^3/t_i], p_m - methane concentration in the captured gas [%], t_i - time unit [year/day/h/min].

Next, the dynamics decrease rate in methane concentration in the captured gas (Q_{pm}) [% / ti] is determined from:

$$Q_{pm} = (p_{pm} - p_{km}) : t_i \tag{2}$$

where: p_{pm} - methane concentration at the beginning of the period under assessment [%], p_{km} - methane concentration at the end of the rated period [%].

Otherwise, it should be followed in the case of the management of methane obtained with the A bore-hole, with a parallel independent exploitation (by another entrepreneur) of the second bore-hole. The specificity of this issue will be presented on the basis of the research and development work [11].

In the presented case, apart from the objective bore-hole A, located in less favourable conditions occurring in the western part of the gas collector, there was bore-hole B, located in more favourable conditions, in the central part of the collector - with higher capacity, which also exploited methane, but by another entrepreneur. In order to determine the performance dynamics of the A bore-hole for the optimal use of the AMM methane deposit for cogeneration engines, several tests were carried out, using an additional flare gas combustion plant, to which methane was supplied through a branch

IOP Conf. Series: Materials Science and Engineering **545** (2019) 012017 doi:10.1088/1757-899X/545/1/012017

mounted at the gas collection site (with a cogeneration aggregate). This ensured a parallel gas collection by the engine working in the aggregate - with a flow of methane to it in the amount of 5 m^3/min - and gas flow to the torch, which imitated the second cogeneration unit. Prior to the tests, the transfer capacity of the A bore-hole was about 880 m^3/h, and the methane concentration in the captured gas was about 61%.

To illustrate the method three incidents influencing the layout of the deposit - gas combustion, were selected from the tests, mainly due to the fact that they enabled to recognize the deposit during unusual phenomena, occurring from time to time. Two incidents were related to an operation of the pressure transfer station, one – causing a necessity of limiting the flow of captured gas through the torch to approx. 840 m^3/h to maintain the minimum pressure needed to operate the generator. The results of these tests indicated that in addition to the incident periods, there were changes regarding:

- the amount of captured gas - from 880, through 890 to 782 m^3/h,
- gas temperature - from 36.56, through 40.6 to 38°C,
- head vacuum - from -8.81, through -8.95 to 8.89 kPa,
- methane concentration - from 61.14, 62.82 to 51.95%.

as shown in Figures 2 and 3.

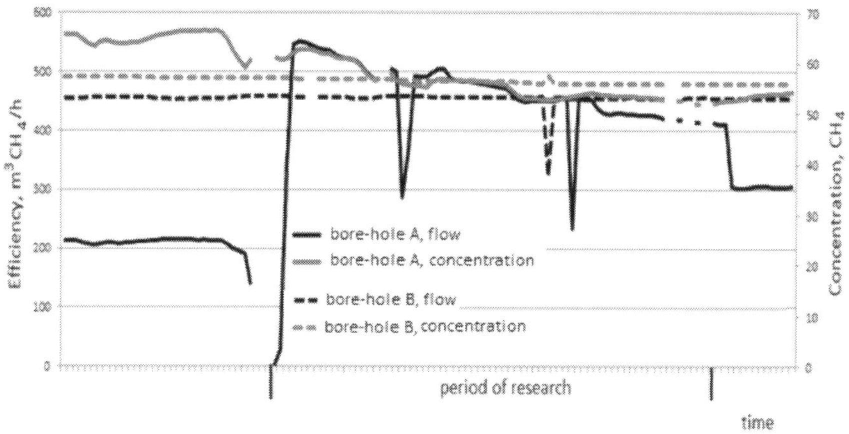

Figure 2. Illustration of changes in methane yield and concentration in bore-holes A and B [11].

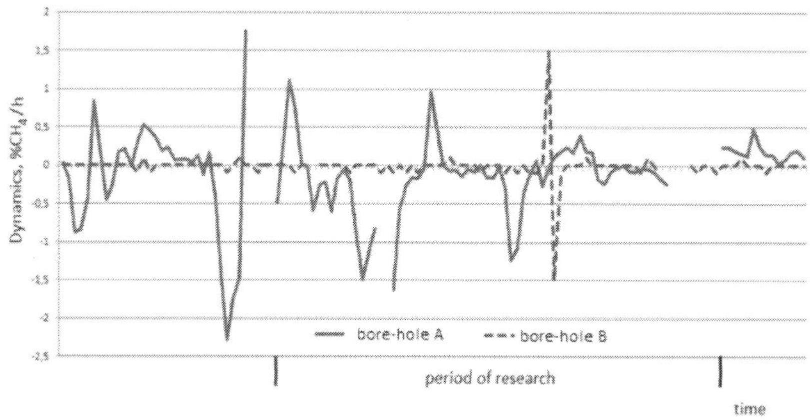

Figure 3. Illustration of the dynamics of changes in the concentration of methane captured by holes A and B during the testing period [11].

The tests, simulating an operation of two cogeneration engines on bore-hole A, showed as follows:

- it is possible to increase the methane extraction with the A bore-hole to the level limited by the pumping station efficiency and limited by the decreasing methane concentration,
- it is impossible to operate the pumping station with maximum efficiency over a longer period of time,
- there was an increase in gas production from bore-hole A:
  - an operation of two cogeneration engines affected the decrease of methane concentration, while the decrease dynamics is lower than the decrease dynamics occurring before and after the experiment,
  - an operation of two cogeneration engines forced the inflow of gas to bore-hole A from a slightly further distance from the bore-hole, mainly from the east, where methane concentration is higher,
  - an operation of two cogeneration engines affected the minimum decrease in the concentration of methane on the B bore-hole,
- the tests confirmed a stabilization of the parameters of the gas extracted after the tests were completed on the bore-hole A in principle within an hour, with the methane concentration stabilizing at a lower level by approx. 2% in relation to the level immediately before the tests,
- an increase in the volume of captured gas on the bore-hole A causes a significant drop in the methane concentration - the dynamics rate of methane concentration drop during the experiment reached the value of 2.4% CH_4/ day, and was 8 times higher than the value of the methane drop indicator for stabilized work for one cogeneration engine (from the period before the tests), amounting to 0.2987% CH_4/day.

These applications were used for planning the exploitation method, which resulted in a selection of a stable, long-term use of AMM gas for energy production.

7. Summary

Undertaking activities, aimed at improving the environmental conditions of the Polish society, results from the diagnosis of the current bad condition of atmospheric air in many cities, which is the result of suspended dust and benzo(a)pyrene from domestic furnaces and local coal-fired boilers.

The actions of the Polish government related to, among others, subsidies for exchanging coal boilers for individual power generation for ecological boilers and gas heating, as well as to an implementation of the Polish Energy Policy until 2040, assume a shift from hard coal and lignite as the basic sources of energy.

In the diversification of energy sources, methane also has a role to play, including the methane obtained from liquidated mines.

Planning of economically justified energy production in cogeneration aggregates requires an identification of the methane deposits in liquidated mines in terms of exploitation resources and deposit dynamics.

A diagnosis of the bed dynamics, requiring research and tests to simulate the AMM gas intake by two production units, enables to choose correctly one of the two methods of reservoir management: stable - long-term or dynamic - short-term, implemented with the use of two or more cogeneration aggregates.

A gradually reduced share of hard coal for energy production will result in a liquidation of subsequent mines, which should increase the share of AMM gas in the pro-ecological energy production.

References

[1] Raport NIK „Ochrona powietrza przed zanieczyszczeniami" 2018 Warszawa
[2] Program rządowy „Czyste powietrze". Ministerstwo Środowiska, Warszawa 2018 *https://www.gov.pl/web/srodowisko/czyste-powietrze – dostępność 02.03.2019 r.*

IOP Conf. Series: Materials Science and Engineering **545** (2019) 012017 doi:10.1088/1757-899X/545/1/012017

[3] Polityka Energetyczna Polski do 2040 roku. Ministerstwo Energii, Warszawa 2018 *https://www.gov.pl/documents/33372/436746/PEP2040_projekt_v12_2018-11-23.pdf/ee3374f4-10c3-5ad8-1843-f58dae119936 – dostępność 01.03.2019 r.*
[4] Lewandowski W M 2007 Proekologiczne odnawialne źródła energii (WNT: Warszawa)
[5] Energia – dokument GUS file:///C:/Users/User/Downloads/zuzycie_paliw_i_nosnikow_energii_w_2017_ r..pdf – dostępność 01.03.2019 r.
[6] Energia 2019 (GUS: Warszawa) *https://stat.gov.pl/files/gfx/portalinformacyjny/ pl/defaultaktualnosci/5485/1/6/1/energia_2018.pdf* – dostępność 01.03.2019 r.
[7] Hadro J I and Wójcik I 2013 Metan pokładów węgla: zasoby i eksploatacja *Przegląd Geologiczny* **61/7**
[8] Zagospodarowanie metanu wg ARP *https://www.cire.pl/item,144354,1,0,0,0,0,0,w-2016-r-kopalnie-wykorzystaly-wiecej-metanu-do-produkcji-energii-elektrycznej-ciepla-i-chlodu.html – dostępność 01.03.2019 r.*
[9] Obwieszczenie Marszałka Sejmu Rzeczypospolitej Polskiej z dnia 16 października 2017 r. w sprawie ogłoszenia jednolitego tekstu ustawy — Prawo geologiczne i górnicze. Dz. U. z 2017 r. poz. 2126.
[10] Trenczek S 2017 The determination of methane resources from liquidated coal mines *Role of Polish coal in the national and European energy sector* IOP Conference Series-Materials Science and Engineering **268** Conference on Role of Polish Coal in the National and European Energy Sector (Andrychów 19-21.06.2017)
[11] Trenczek S and Krause E 2017 Optymalizacja wydobycia i zagospodarowania metanu z odwiertu w kontekście wyników przeprowadzonych badań specjalistycznych złoża *Dokumentacja usługi naukowo-badawczej* Katowice (not published)

IOP Conf. Series: Materials Science and Engineering **545** (2019) 012018 doi:10.1088/1757-899X/545/1/012018

Supporting development of suspended underground monorails using virtual prototyping techniques

K Szewerda

KOMAG Institute of Mining Technology, Laboratory of Modelling Methods and Ergonomics, Pszczyńska 37, 44-101 Gliwice, Poland

kszewerda@komag.eu

Abstract. Self-propelled suspended railways are one of the main means of auxiliary transport in underground coal mines. Their dynamic development and increase of their widespread use in Poland took place at the turn of the 20th and 21st centuries. Due to their advantages which include among others no need to maintain tracks on the floor and transport safety. The suspended railway is used instead of other means of transport. Individual subassemblies of the suspended monorails are still under development. Improvement of safety and increase of the work comfort of the operator and conveyed persons are among the objectives of manufacturers of these machines. A possibility of increasing the speed of suspended monorails, especially when conveying people, associated with a continuous extension of the access roads to coal mining faces, is also an important aspect. Increasing the speed will enable a more efficient used employees' working time. This article presents an example of using one of virtual prototyping methods to support the modernization process of existing solutions or to support designing of new machines. Use of this method for a modification of the operator's suspended cabin in a suspended monorail is described.

1. Introduction

The transport system in underground hard coal mines includes the main haulage means, connected with the transport the run of mine to the surface as well as auxiliary means of transport to convey people and to transport materials and equipment to the workplace. The floor-mounted railway, rope-driven suspended monorails and the suspended monorails with their own drive are used in the auxiliary transportation systems. A widespread use of self-propelled suspended monorails increased significantly at the end of the twentieth century. This is confirmed by a rapid increase in the length of suspended monorails routes in the Polish hard coal mines. The suspended monorails gradually replace other auxiliary means of transport. An intensive development and numerous implementations of the suspended self-propelled monorails result from their advantages. Their basic advantages are as follows: no need to keep the track on the floor or a much better mobility as well as increased safety compared to rope-driven suspended monorail [[1], [2], [3]]. However, self-propelled suspended monorails are still under development and there is a lot of intensive research work to be done in this area. Among the objectives that the manufacturers of these machines strive for is an improvement of safety or increase of the operator and conveyed personnel's comfort. A possibility of increasing the maximum speed of suspended monorail, especially while conveying of personnel is also an important aspect. It is connected with a continuous extension of the routes length to work places. An increase of

IOP Conf. Series: Materials Science and Engineering **545** (2019) 012018 doi:10.1088/1757-899X/545/1/012018

the speed limit will enable a more efficient use of employees' work time [[1], [4], [5], [6], [7], [8], [9], [10]].

Conceptual work on new solutions used in suspended monorails designs or research work aiming at an improvement of already existing designs can be supported by use of virtual prototyping techniques [[11], [12]]. An analysis of kinematics and dynamics of multibody systems (MBS) is one of these methods. Its use enables, among others, to speed up the designing process of new solutions or a selection of correct parameters of sub-assemblies of suspended monorails. An analysis of stiffness impact of elastic-damping components, used in the suspension of the operator's cabin on the monorail operator's overload, is one of the examples of using the MBS method. By changing the properties of these components, the operator's overload, especially the one acting in the vertical axis, can be significantly reduced or even eliminated. Correctly selected parameters of elastic-damper components improve the operator's work comfort, which is especially important in terms of a possibility of increasing speed limits of suspended monorails. The results, presented in the article, are developed within the INESI project (Increase Efficiency and Safety Improvement in Underground Mining Transportation Routes), conducted by the international consortium, coordinated by the KOMAG Institute of Mining Technology [[1], [4], [8]].

2. Analysis of the operator's cabin suspension

2.1. Verification of the computational model

A preparation and a verification of the computational model are required to conduct numerical simulations and analysis of the stiffness impact of elastic-damping components to overloads acting on the operators of suspended monorails. Figure 1 shows a simplified computational model used to simulate an impact of the stiffness coefficient change of elastic components of the operator's cabin suspension to overloads, acting on the operator during his work. This model consists of the operator's cabin (1) of 450 kg weight, the trolley of the operator's cabin (2) of 208 kg weight, a toothed gear drive (3) of 640 kg weight and the suspended monorails which include the rails of 2 m length (4) and its slings (5). The operator's cabin is connected to the trolley by four elastic-damping components (E1 – E4).

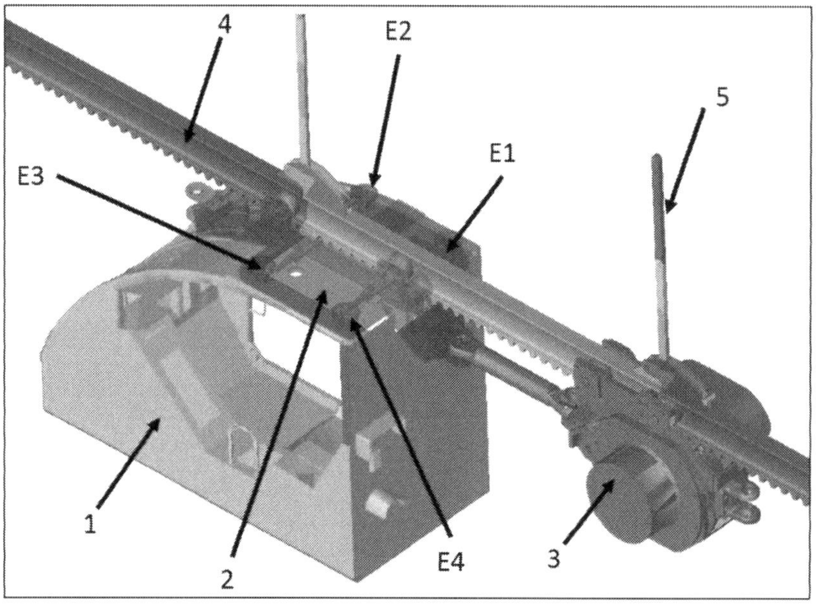

Figure 1. Computational model of the operator's cabin [4].

IOP Conf. Series: Materials Science and Engineering **545** (2019) 012018 doi:10.1088/1757-899X/545/1/012018

The stiffness coefficient of the elastic components E1 - E4 has been selected in such a way that it corresponds to the components currently used in the construction of the operator's cabin. Figure 2 shows the method for the computational model adjustment and verification.

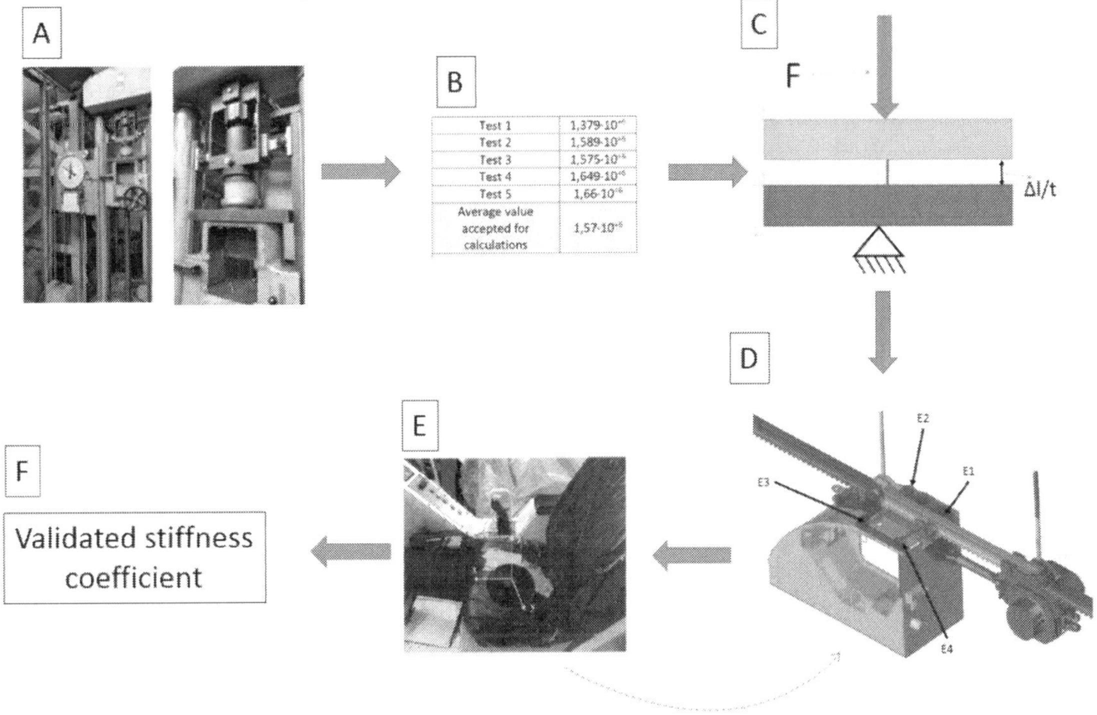

Figure 2. The method for the computational model adjustment and verification [4].

The first step (Figure 2 A) consisted in the compression test of the elastic suspension component made of the currently used material. These tests were carried out on a strength testing machine. The compressive force and the deflection of the tested component were recorded during five tests. The test stand, on which the compression tests were carried out, is shown in Figure 3.

Figure 3. Test stand - the strength testing machine [4].

On the basis of the results of the carried out tests, the stiffness coefficient of the tested component was calculated in relation to each compression test of the elastic suspension component. Then the average stiffness coefficient was calculated (Figure 2 B). The values calculated on the basis of the tests are shown in Table 1.

Table 1. The results of elastic component compression tests of the operator's cabin suspension [4].

No.	Stiffness coefficient [N/m]
Test 1	$1.379 \cdot 10^{+6}$
Test 2	$1.589 \cdot 10^{+6}$
Test 3	$1.575 \cdot 10^{+6}$
Test 4	$1.649 \cdot 10^{+6}$
Test 5	$1.66 \cdot 10^{+6}$
Average value	$1.57 \cdot 10^{+6}$

Next, a simplified computational model of the test stand was prepared. It consisted of two jaws, the lower one was fixed and the upper one exerted force on the elastic-damping component which replaced the tested elastic component (Figure 2 C). The average stiffness coefficient of the elastic-damping component, calculated on the basis of the previous step of the stand test, was introduced. A value of the compressive force increase was defined in the same way as for the test stand. A deflection of the elastic-damping component of the set stiffness coefficient was recorded during simulations. This step was conducted to verify the correct operation of the elastic - damping component. The deflection values of the elastic damping component, obtained in the model as a result of the set compressive force, were compared with the deflections of the tested suspension component obtained on the test stand. Figure 4 shows a simplified computational model of a test stand and two sample charts showing the comparison of the results calculated by numerical simulation and recorded on the test stand.

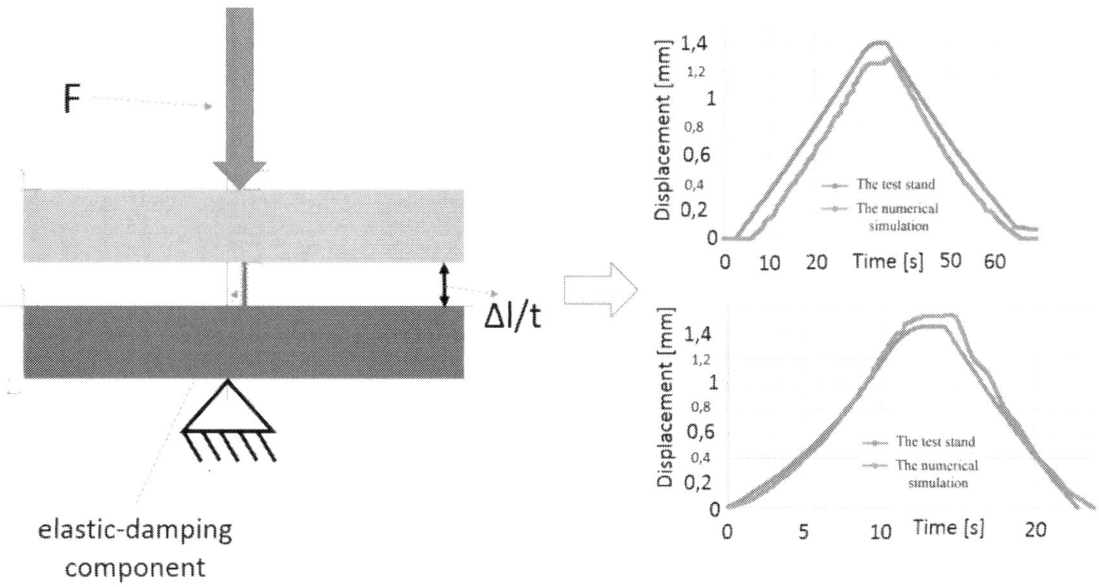

Figure 4. The calculation model for verification of the elastic - damping component and test results [4].

After verifying the correct operation of the elastic-damping component as a flexible component of the operator's cabin suspension, the determined stiffness coefficient was introduced to the computational model of the operator's cabin (Fig. 2 D). This value was used in all elastic components of the suspension E1 – E4. Then, a travel of the operator's cabin on a straight section of the suspended monorail, consisting of 2-meter long rails (Fig. 5) was simulated. The cabin speed during simulations was equal to 2 m/s. The acceleration acting on the operator in the cabin was recorded during the cabin movement simulation.

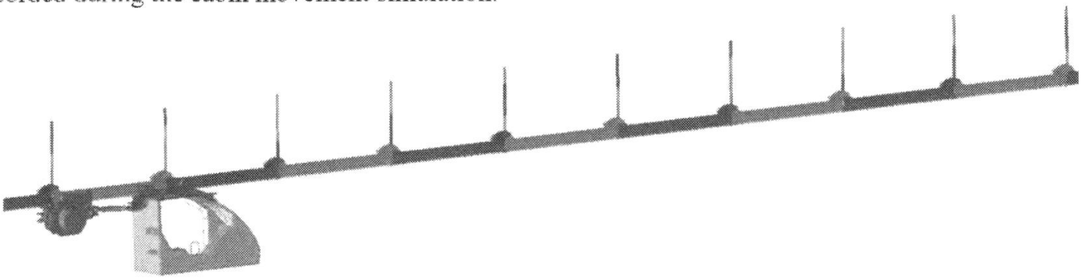

Figure 5. Computational model with a straight monorail segment [4].

The next step in a verification of the operator's cabin computational model was to compare the overloads acting on the operator calculated by numerical simulations with the values recorded during the vibration tests in the real cabin (Figure 2 E). The tests were carried out on the test track by the machine manufacturer. The overloads were recorded both in the operator's front and end cabins of the real object. The analyzed components of the operator's cabin suspension were designed to minimize the overload impact acting on the operator mainly in the vertical axis. This is the reason why in the process of the computational model verification an attention was paid to this type of overloads.

In the result of comparing values measured on the real object and determined in the numerical calculations, the stiffness coefficient of the E1 – E4 components was corrected (Figure 2 F). A repetition of the step marked with the letter E in Figure 2 may be required for more accurate imaging of the real object by the computational model to correct the stiffness coefficients. After the correction, in further calculations, the stiffness coefficient of components E1 – E4 was assumed to be equal to $1.64 \cdot 10^{+6}$ N/m. Table 2 presents the comparison of overloads acting in the vertical axis, recorded on the real object and calculated on the cabin computational model after the correction of the stiffness coefficient of E1 – E4 components.

Table 2. Comparison of the vertical axis overloads calculated in numerical simulation with those recorded on the real object [4].

No.	Stiffness coefficient [m/s^2]
Real measurement	Cabin 1: 0.66 – 0,69
	Cabin 2: 0.62 – 0,65
Numerical simulation	0.6326

On the basis of the presented comparison of overload values acting on the operator, it is assumed that the computational model of the operator's cabin is verified and can be used in numerical simulations and analyses of the stiffness impact of the cabin suspension components on the operator's overload.

2.2. Impact analysis of change in the stiffness coefficient of the suspension components on the operator's overloads

Monorail connections in the suspended monorail track should allow for the rails deflection against each other in the vertical axis by less than 3° in both directions. A change in the position of rails against each other can affect the operator's overloads, when passing through these connections. Due to this fact, the computational model of the operator's cabin was modified by introducing a possibility of changing a deflection of two rail sections on the track on which the cabin moves. The computational model with the modified track is shown in Figure 6.

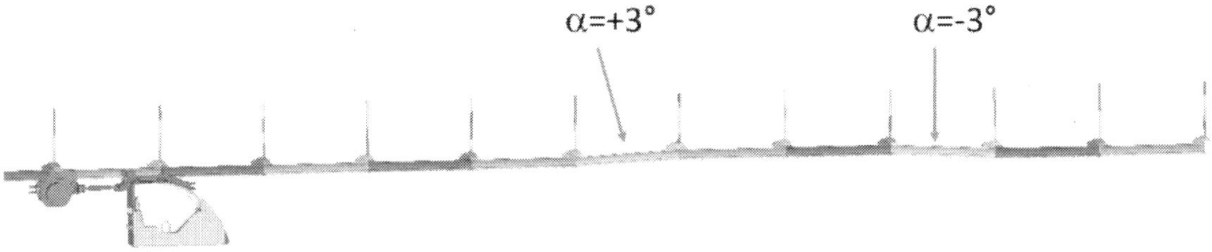

Figure 6. The computational model of the operator's cabin with the modified track [4].

The computational model prepared in this way was used in numerical simulations. And then, basing on the obtained results, an impact of the stiffness coefficient of the suspension components on values of the overloads acting on the operator in the vertical axis, was analyzed. For this purpose, within the INESI project research work, three materials of different stiffness coefficients were selected.

The components of the operator's cabin suspension were made of these materials. Similarly as for the suspension components made of currently used material, the stiffness coefficient was determined for each of the selected materials. Table 3 presents the determined stiffness coefficients of the analyzed materials.

Table 3. Stiffness coefficients accepted for numerical calculations [4].

Material	Stiffness coefficient [N/m]
Current material	$1.64 \cdot 10^{+6}$
Material A	$2.122 \cdot 10^{+6}$
Material B	$9 \cdot 10^{+5}$
Material C	$4.72 \cdot 10^{+5}$

Then, five numerical simulations of the operator's cabin movement on the modified track were conducted. The simulations were performed assuming the stiffness coefficient of the currently used suspension, i.e. the stiffness coefficient determined for the materials A, B and C and for the operator's cabin fixed to the trolley. The overload which affected the operator during the motion was recorded in each simulation. Figure 7 shows the recorded overloads experienced in the vertical axis in the case of the analyzed materials.

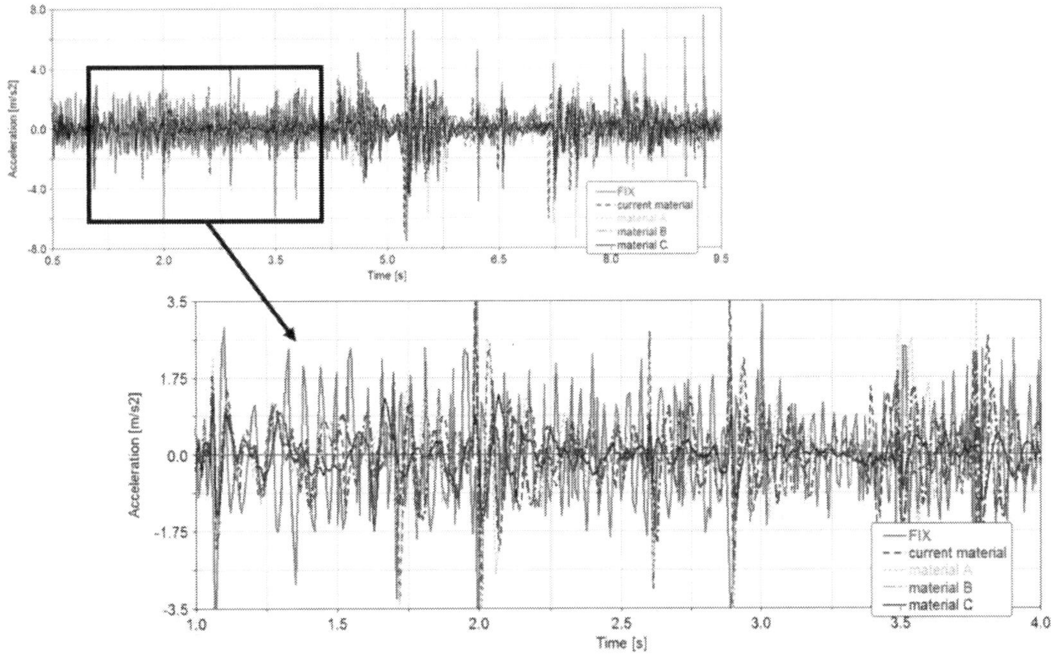

Figure 7. The overloads acting on the operator in the vertical axis [4].

A deflection of the components of operator's cabin suspension was another parameter recorded during the simulation. A deflection of the suspension (E1) elastic component of in relation to the analyzed materials is shown in Figure 8.

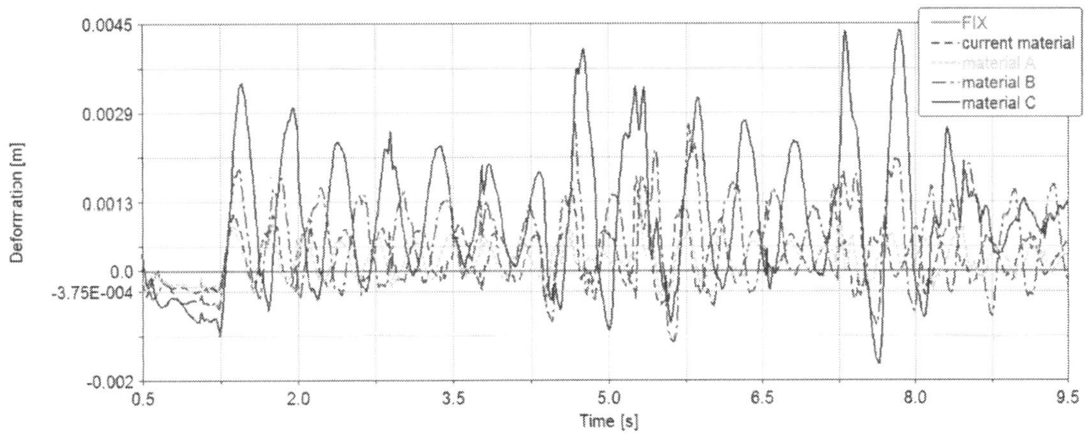

Figure 8. Deformation of the elastic component E1 of the operator's cabin suspension [4].

Figure 9 shows a comparison of the numerical calculations results of in relation to the material with the highest (material A) and the smallest (material C) stiffness coefficients of the cabin suspension elastic component.

The overloads affecting the operator in the cabin, in the vertical axis, are presented in Table 4. The table shows the maximum, minimum and RMS value of overloads determined by numerical simulations in relation to four selected materials used for a manufacture of the cabin suspension elastic components as well as in relation to the fixed joint between the operator's cabin and the trolley.

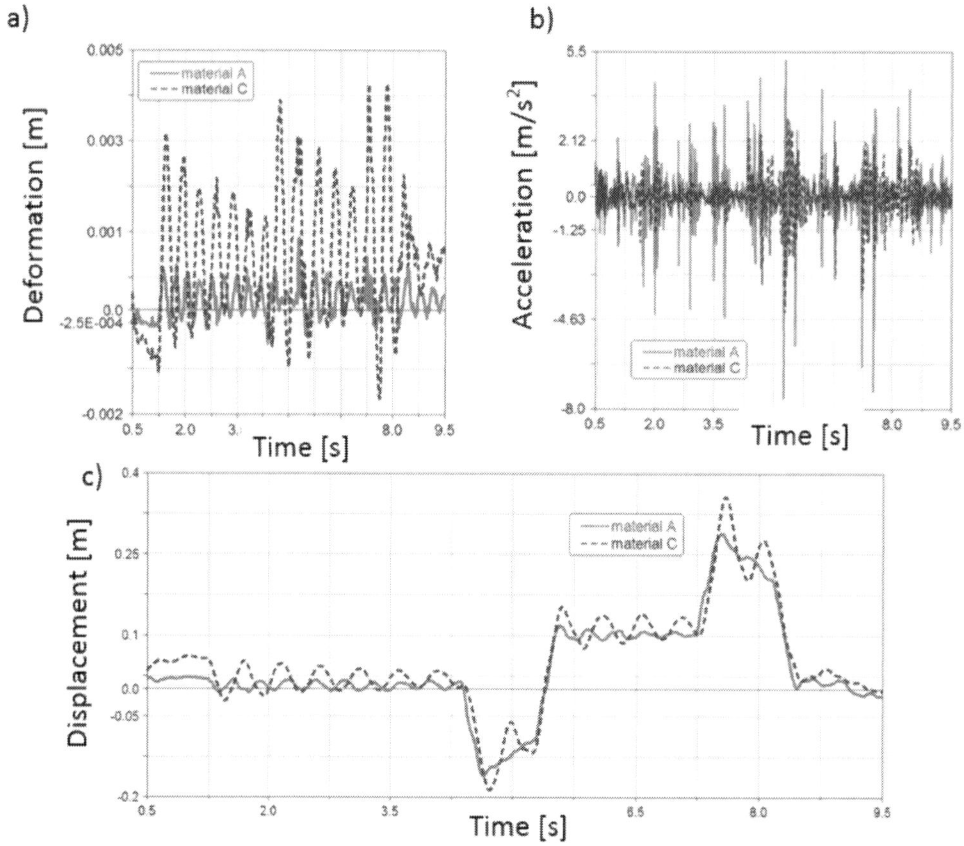

Figure 9. Comparison of simulation results for the material with the lowest and the highest stiffness coefficients a) deformation of the elastic suspension component, b) overload acting on the operator, c) displacement of the operator's cabin in the vertical axis [4].

Table 4. Maximum, minimum and RMS values of overloads affecting the operator in relation to different materials used for elastic components of the cabin suspension [4].

Material	Maximum overload [m/s^2]	Minimum overload [m/s^2]	RMS value of overload [m/s^2]
Fixed joint between the cabin and the trolley	10.0523	-7.5171	1.3431
Current material	5.0549	-7.0158	0.9692
Material A	5.1997	-7.6237	1.0958
Material B	3.7267	-4.3735	0.869
Material C	2.9748	-4.3695	0.7288

Basing on an analysis of the results obtained in numerical simulations, it can be concluded that the stiffness of elastic components used in the operator's cabin suspension has a big impact on the overloads affecting the operator during his work. The fixed joint between the operator's cabin and the trolley resulted in the biggest overloads affecting the operator's body. It was observed that the material currently used for a manufacture of elastic components, despite the fact that it reduces overloads, is not an optimal solution. Material A from the analyzed materials turned out to be too stiff and thus its use causes an increase in the operator's overload in relation to the currently used material. The most favourable overloads acting on the operator were observed when using the material with the smallest stiffness coefficient, i.e. Material C. Its use resulted in the biggest deflection of the elastic components in the suspension of the operator's cabin during the motion on the track test segment. It also resulted in the biggest displacement of the operator's cabin and the smallest overloads acting on the operator.

3. Conclusions

The article presents an example of using the MBS numerical simulations in analyzes aimed at a development of suspended monorails solutions. Basing on simulations, the impact of changing the elastic component stiffness coefficient of the operator's cabin suspension was analyzed, concluding that from the materials selected by the manufacturer, the use of type C material is best to minimize the overloads to the operator. In this way, a change of the material used for the elastic components will improve the operator's work comfort.

The computational model was limited only to the operator's cabin of the suspended monorail. However, it is possible to expand the computational model in such a way that it can represent any component of the suspended monorail. The current computational model, developed by the author, also includes a cabin for personnel transportation. Similarly to the method of the analysis described in the article, it is possible to assess the overloads affecting the conveyed personnel. In addition, the use of computer simulations enables an analysis of overloads that affect both the operator and the miners carried in the passenger cabin, not only during normal operation of the suspended monorail but also in emergency situations such as emergency braking. Such simulations enable to estimate hazards in different situations, such as emergency braking while going upward and downward or emergency braking at different speeds. Such a comprehensive analysis enables to develop innovative systems aimed at increasing the safety of personnel conveyed by suspended monorail systems.

In addition, the use of MBS numerical simulations enables an analysis of forces and torques acting at any place of the model. This feature can be used, for example, to analyze the overloads to suspended track suspensions or forces acting in the rail joints. Such knowledge contributes to a development of suspended monorail transportation systems, by modifying the existing designs in order to increase the transportation capacity of the suspended monorail or to analyze a possibility of increasing the maximum speed limit of the suspended monorails. The results of the MBS simulations can also form be boundary conditions for strength calculations of each subassembly of the suspended monorails.

Acknowledgments. The paper was written as part of the work carried out under the European project INESI, "Increase efficiency and safety improvement in underground mining transportation routes". This project is financially supported by the Research Fund for Coal and Steel under the Grant Agreement No 754169. Calculations were carried out at the Academic Computer Centre in Gdańsk, Poland.

References

[1] Tokarczyk J 2017 *Methodology for identifying the selected mechanical hazards in auxiliary transport of underground mines* In: Monograph vol. **52**, Institute of Mining Technology KOMAG, Gliwice

[2] Szewerda K 2014 Metoda analizy kinematyki i dynamiki układów wieloczłonowych do identyfikacji sił w modułowych zestawach nośnych *Masz. Gór.* **3** pp. 3-10, ISSN 0209-3693

[3] MINTOS Project: Improving Mining Transport Reliability. RFCS Coal RTD Programme, Contract No. RCR-CT-2007-00003. Gliwice, 2007÷2010 (not published).

[4] INESI European Project: Increase Efficiency and Safety Improvement in Underground Mining Transportation Routes. RFCS, Contract No. 754169 (2017-2020)

[5] Pieczora E Suffner H 2017 Rozwój napędów dołowych kolejek podwieszonych. *Masz. Gór.* 2017 **3** pp. 44-57, ISSN 2450-9442.

[6] Zasadni W et al. 2015 Możliwości zwiększenia prędkości jazdy kolejkami podwieszonymi z napędem własnym. Conference proceedings :XVII Konferencja: Problemy bezpieczeństwa i ochrony zdrowia w polskim górnictwie. Zawiercie

[7] Budniok T et al. 2014 Analiza możliwości zwiększenia prędkości jazdy ludzi kolejkami podwieszonymi z napędem własnym. Conference Proceedings: XXII Międzynarodowa Konferencja Trwałość Elementów i Węzłów Konstrukcyjnych Maszyn Górniczych TEMAG 2014. Gliwice-Ustroń

[8] Tokarczyk J Dudek M Szewerda K and Turewicz A 2014 Stan i kierunki rozwoju systemu wspomagania prac inżynierskich dla pomocniczego transportu kopalnianego Safe Trans Design (STD) *Maszyny. Górnicze* **2**

[9] Lesiak K Brzeżański M 2018 Concept of the exhaust system for diesel engines used in underground mining. Conference proceedings: KONMOT 2018, International Automotive Conference, Kraków, September 13th-14th, 2018 pp. 1-6, ISSN 1757-899X

[10] Polnik B 2018 An Innovative Power Supply System Dedicated for Roadheading Mining Machines. (ECS Transactions, 2018 No. 1). Conference proceedings: ABAF 2018, 19th International Conference on Advanced Batteries, Accumulators and Fuel Cells, Brno, 26-29 August 2018 349-362, ISSN 1938-6737.

[11] Tokarczyk J 2015 Method for virtual prototyping of cabins of mining machines operators *Arch. Min. Sci* **60** (1) pp. 329-340.

[12] Tokarczyk J 2016 Method for identification of results of dynamic overloads in assessment of safety use of the mine auxiliary transportation system *Arch. Min. Sci.* **61** (4) pp. 765-777.

IOP Conf. Series: Materials Science and Engineering **545** (2019) 012019 doi:10.1088/1757-899X/545/1/012019

Hard rock mining – cutting or disk tools

K Kotwica

AGH University of Science and Technology, Faculty of Mechanical Engineering and Robotics, Al. Mickiewicza 30, 30-059 Cracow, Poland

kotwica@ahg.edu.pl

Abstract. The article presents problems related to the mining of compact and very hard rocks with cutting tools, mainly with tangential-rotary picks. The currently used disc tools, their parameters and the way of working are described. The disadvantages of selected solutions and related obstacles to their wider use in mining have been presented. The directions of further development of disk tools were presented, allowing their use, for example, in mining heads of longwall shearers and roadheaders.

1. Introduction

In underground mines, both in Poland and in the world, one of the most commonly used methods of hard rocks mining, in the process of row minerals excavation and drilling of exploratory and opening-out headings or tunnels is the mechanical method. It allows you to obtain large mining capacity or drilling speed, but also generates a number of hazards or restrictions, depending of used mining tools.

The roadways, especially opening-out headings, are currently excavated in rocks with very unfavorable parameters. This applies above all to the high strength of the rock massive on uniaxial compression, which in many cases exceeds 120 MPa and its structure. Increasingly, it encounters rocks that have a homogeneous structure, which results in their mining encounters serious problems, especially when using mechanical methods. No less important factor is the content of minerals and inclusions in the rocks, causing rapid abrasion and wear of mining tools and in the case of inclusions, for example spherosiderites, the occurrence of sparks during mining process [1]. The article presents the applied methods of mechanical mining of rocks, with particular emphasis on cutting and disc tools as well as hazards (dust, spark, wear) and limitations (rock compactness, machine dimensions) occurring in these methods.

2. Machines and tools used in mechanical mining of rocks

Currently, the most commonly used machines for roadway and tunnel excavation in hard rocks are roadheaders, continuous miners, multi-organ combines, special combines and TBM (tunnel boring machines) with full face excavation. Examples of such machines are shown in figure 1. In the case of roadheaders and continuous miners, as well as multi-organ combines, cutting tools are used, mainly tangential-rotary picks. In the case of special combines and full face excavation machines type TBM, symmetrical or asymmetrical disk tools are used.

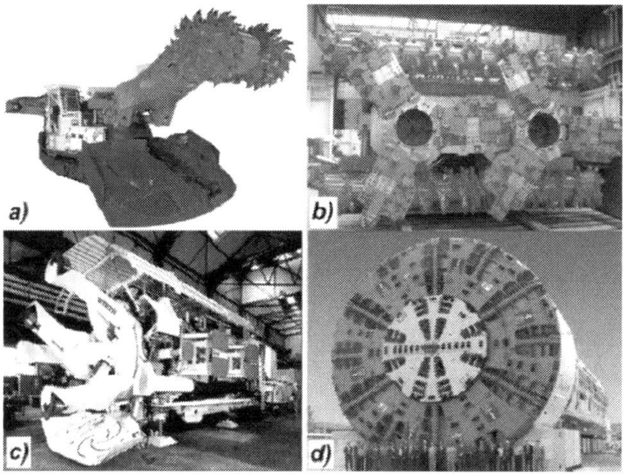

Figure 1. View of: a) roadheader type MR 520 made by Sandvik, b) multi-organ combine type MF 420 made by Sandvik, c) special combine applying undercutting mining method made by Wirth, d) TBM shield full face excavation machine made by Herrenknecht [2,3,4].

2.1. Cutting tools - working principle, advantages and disadvantages

Two types of cutting tools are used on milling heads of mining combines - radial and tangential-rotary picks. On the edges of these picks are mounted sintered carbide inserts or posts. In spite of this, and correctly selected resting and movement cutting angles, these picks are exposed to rapid friction wear, even when soft rocks are mined. However, when mining compact rocks, working part of radial pick can be damaged due to the high bending forces [5].

These inconveniences have caused that nowadays on shearers or roadheaders for hard rocks mining, mainly tangential-rotary picks are used. Thanks to their design and the method of fixing, there are able to rotate in tool holders. These tools are more predisposed to mining rocks, even very hard ones, but they need to maintain proper parameters of their work. They are currently used on mining heads of almost all roadheaders and most of the longwall shearers. The view of the structural solutions of radial and tangential-rotary picks is shown in figure 2 [1,5].

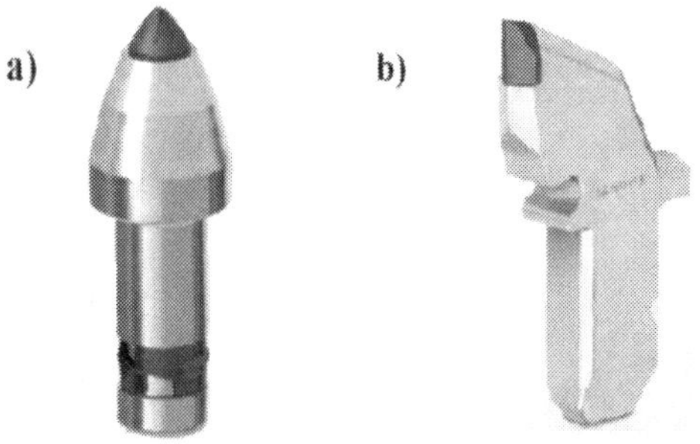

Figure 2. Types of cutting tools used in cutting or milling: a) rotary-tangential pick, b) radial pick [4].

However, the disadvantage of the method of mining rocks by milling is the generation of dust and limitation of the use of this method, associated with the upper limit of rock strength to uniaxial compression or high compactness of these rocks. In connection with the high abrasiveness of such rocks, this results in excessive wear of cutting picks and a decrease in drilling speed. Excessive wear of the tool edge also leads to increased dustiness and the risk of explosion of mine gases, mainly methane, as a result of frictional sparking (figure 3). This problem occurs especially with radial picks. Even the use of rotary-tangential pick reduces this problem only to a small extent.

Figure 3. View of frictional sparking and dustiness during cutting of a rock sample with a rotary-tangential pick and wear of pick wedge.

When mining hardly mineable rocks, the increased wear of tools and the generated dust hazards and gas explosion often result that the process of rock mining is economical unprofessionally (low efficiency and high energy consumption of the mining process, high wear of tools). Work is underway to develop more durable cutting tools, however, even new-generation rotary-tangential pick are not always able to provide the required mining parameters [1, 5].

The new solution of special tools, which can replace traditional tangential-rotary picks was developed in AGH UST. This tools are adapted to attach in standard holders of tangential rotary pick which are commonly used for hard rock cutting, whereas design solution of the picks working part is different than those designed for the standard tangential rotary picks [6]. The new pick is shown in figure 4. The cutting segment is bell or crown-shaped, and the working part is armed at the circumference with eight cone-shaped sintered carbide inserts. Such design solution, instead of traditional cutting with use of standard tangential rotary picks, should allow loosening rock fragments as a result of point pressures exerted by individual sintered carbide inserts. Moreover, non-uniform load of individual sintered carbide inserts should cause increase the tool rotary speed in the holder, at minimal side deflection angle of the cutting tool. After manufacturing of the new tool prototype, preliminary tests at a special laboratory stand were carried out.

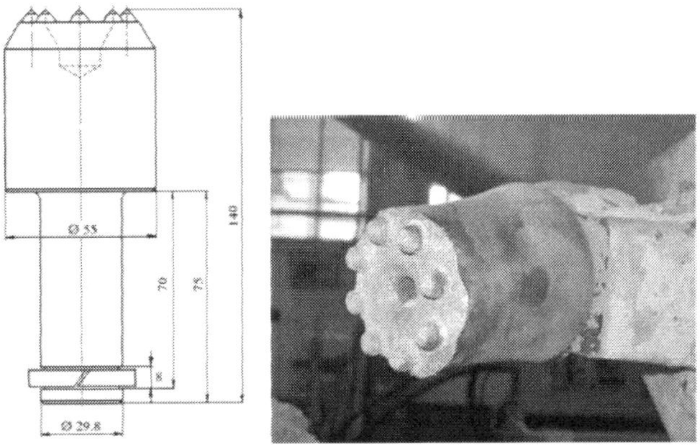

Figure 4. Scheme of the new type of the crown pick and view of new crown pick with 8 carbide inserts with diameter 8 mm after the test [6].

The results of preliminary researches has been satisfactory. Significant rotations and very small wear of new tool were observed. Figure 4 presents the view of this crown pick after sample mining on the total

IOP Conf. Series: Materials Science and Engineering **545** (2019) 012019 doi:10.1088/1757-899X/545/1/012019

distance of 1200 m. New type of the tool has been applied in the new solution of roadheader mining head. The mining head of the R-130 roadheader, equipped with new crown picks, was tested in Iłża limestone opencast mine (figure 5a). Working conditions were comparable to the standard mining head with tangential rotary picks operation regime. There were no problems with the head work. During 5 hours of work, a comparable quantity of output (to a standard mining head) was excavated. The difference was only in dimensions of the grain size of the output. The output was homogeneous and fine. No major signs of wear of new tools were observed. The view of one of the new pick after the tests is shown in figure 5b. In order to confirm the effectiveness of these tools, however, research in real conditions in an underground mine is required.

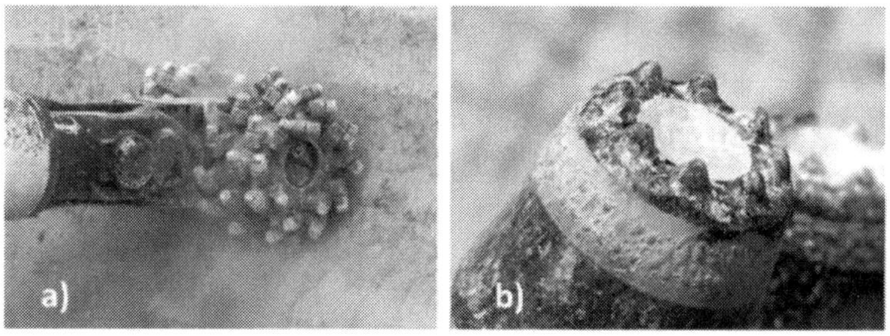

Figure 5. View of the mining head with new crown picks: a) – during the tests in Iłża limestone opencast mine, b) – new crown picks after the tests [6].

2.2. Disk tools - working principle, advantages and disadvantages

The second of the most commonly used methods in the process of mechanical mining of hard rocks is mining by means of static crushing, carried out with the help of wheel tools called disks. Disk tools used for this mining method can be symmetrical or asymmetric. This method consists in pushing the disc edge into a rock massive with the normal force perpendicular to its surface [1]. As a result of this force, the uniaxial compressive strength of the rock is locally exceeded, rock massive is crushed and disk penetrates into the depth g (figure 6). At the same time, the rolling force is applied to the holder, causing the tool to move along the surface of the mined rock. Rotary assembly the disk in the holder allows the disk to rotate around the axis. The ability to rotate the tool in the head reduces the friction forces, which reduces energy loss and generates less toll edge wear and dustiness, compared to cutting tools. In addition, the rotation of the cutting disc causes short contact of the edge section with the rock, which greatly facilitates the removal of heat emitted on the edge of the disk and ensures very long durability of the tool. The typical construction of disk tool consists of shaft, seal, hub, split ring and replaceable cutting ring is shown in figure 7. There are three types of symmetric disk tools: smooth disks, disks armed with carbide insert and toothed discs, in one and double or even triple blades versions. However, the first two types are most often used, which in the view are shown in figure 8. Their diameters can be even up to 500 mm [3,7,8].

TBM disk tools are mounted on the TBM mining head in a certain pattern and cover the entire face. They are exposed to different and sometimes even extreme ground conditions during tunneling, so selecting tools with high quality, reliability and long lifetime is important and critical [7,8]

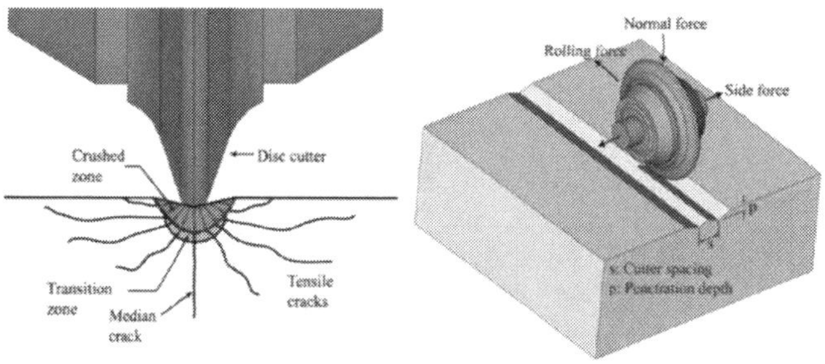

Figure 6. The work principle of a single symmetric disk [3].

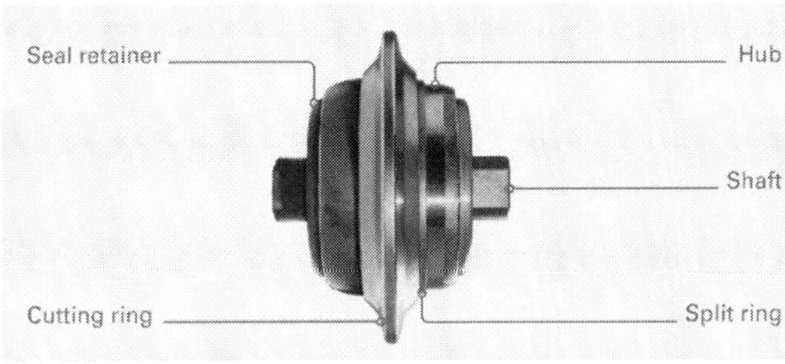

Figure 7. The typical construction of disk tool [3].

Figure 8. View of smooth disk and disk armed with carbide insert in one and double blades versions [1].

Smooth single blade disk tool can works without visible wear even up to few kilometers of roadway or tunnel advance. In comparison cutting tool lifetime is not more than several cubic meter of wining per piece.

The disadvantage of mining by static crushing is the need to ensure a high pressure force of the tool. Destruction of the stone rock structure occurs as a result of exceeding its uniaxial compression strength. According to the available data, the value of the pressure force per disk tool shown in figure 8 can be as high as 300 kN. In the case of a large number of disk tools on the mining head, this gives a very large value of the total pressure force (depending on the diameter of the body up to 25,000 kN) and many times more value of side expanded force to ensure the stability of TBM work. The mining machine must transfer the reactions coming from the mining head and stabilizing struts. The occurrence of large values of reaction forces occures large mass (up to 3,500 Mg) and dimensions of the machine (the length of the TBM reaching up to 400 meters). The view of the mining head and TBM from Robbins is shown in figure 9. This results in the limited use of this type of machinery for the drilling of excavations with a very large runnout, where very large costs related to their production and construction in the mining excavation can be paid back [1,7,8].

Another untypical solution of a milling machine, which uses a principle of mining by static discs and leaning out the boom in a horizontal plane, is a continuous miner made by ROBBINS, which is shown in figure 9. The mining head is constructed in a form of a larger diameter plate with mining disks mounted with bearings around the perimeter. This unit is mounted on an arm, inside of which there is an engine whose power puts the unit into a rotational motion. The arm's leaning out movement makes the advance motion in a vertical plane. This solution enables excavation of roadway in cross-section similar to rectangle. View of roadway obtained by this machine is shown in figure 10 [6].

A clear drawback of the static crumpling method is necessity to provide large pressure forces to the disks. Whereas, asymmetrical disk tools have been used as mining tools for longwall shearers to increase output of a large size grade. Conducted industrial tests demonstrated usefulness of such equipment for obtaining higher product graining. Asymmetrical disk tools are applied in the technique of mechanical mining not only as crushing devices but also in the undercutting method.

Figure 9. View of mining head with disk tools and TBM complex manufactured by Robbins [6].

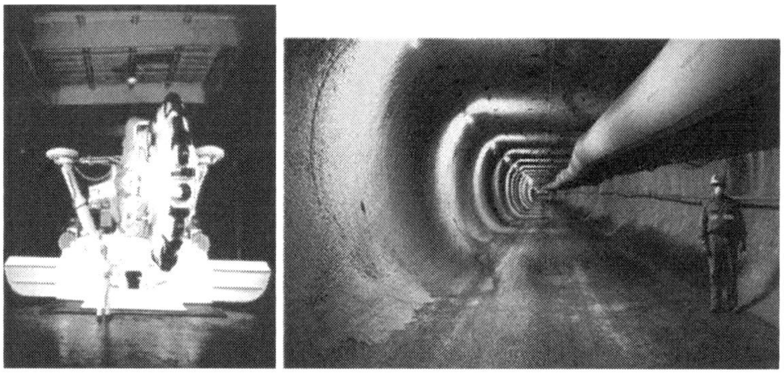

Figure 10. View of Mobile Miner MM130 made by Robbins and excavation performed by this machine in Broken Hill Mine [6].

The principle of the undercutting method is mining a rock by its cutting off towards a free space. A disk tool affects the rock tangentially to the surface of the mined body, similarly as for cutting tools, but the difference is that here it uses the disk rolling movement which efficiently eliminates sliding friction in favor of rolling friction. Figure 11 presents a diagram of mining with this method. Application of disk tools in that way lowers energy consumption and pressure force. It allows designing mining machines with respectively lower energy demands and lower requirements concerning stability than in case of machines equipped with classical disks, operating perpendicularly towards the surface of a mined body [9].

In the case of this method, however, there are strongly varying lateral forces on the edges of disk tools. This is the reason for the difficulty in properly taking over the reaction on the disk tool holders and their bearing. Therefore, after the initial research of a machine developed by Wirth, using this method, equipped with four swinging and rotating arms and ending with positive results at the end of the twentieth century, the further development of machines using the new method of mining rock was stopped [9]. The view of the Wirth machine used undercutting method and the effects of its work is shown in figure 12.

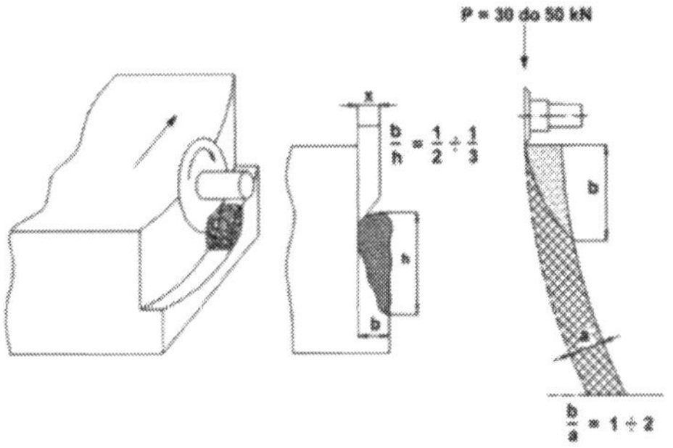

Figure 11. Diagram of the undercutting method mining principle [9].

Figure 12. view of the Wirth machine used undercutting method and the effects of its work [9].

3. Future of disk tools application

For many years, work has been carried out on the development of new construction solutions for mining heads of roadheaders or mining methods that allow effective mining of hardly workable rocks. One of such solutions is a mining head, developed and tested at the Colorado School of Mines [1]. To reduce the values of mining forces, symmetrical mini-discs with a diameter of 125 mm were used. The biggest problem is the bearing of disks, connected with their heavy load. Several solutions for the bearing of tools and a technical design of the body were developed, on the basis of which the prototype was made. The mining head model, along with the arrangement diagram of the tools on its surface, is shown in figure 13. The results of the trials were satisfactory, but no detailed data on this subject.

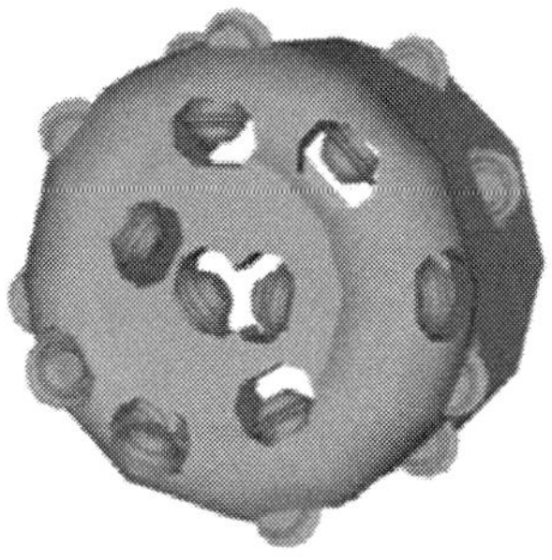

Figure 13. Spatial model of mining head with symmetrical mini disk tools [1].

There were other concepts of use on the mining heads mainly asymmetrical disks, in which the basic idea was to use the disk as a chipping tool. Thanks to this, energy consumption and the value of the pressure force are smaller, which gives the possibility of constructing a mining machine with lower energy parameters, smaller requirements due to stability criterion, as in the case of classic disks operating perpendicular to the rock surface. In the case of machines using this technique, inter alia, by Aker Wirth and Atlas Copco, this

IOP Conf. Series: Materials Science and Engineering **545** (2019) 012019 doi:10.1088/1757-899X/545/1/012019

method showed full usefulness [2, 10]. The only drawback was the complex way of machine control and the large value of reaction forces operating on the machine. View of the Aker Wirth MTM 6 machine developed for the Rio Tinto Company as part of the "Mine of the future" project for the new Resolution copper mines in Arizona as well as Oyu Tolgoi copper and gold mine in Mongolia and the Atlas Copco alternate solution of MMM machine (Modular Mining Machine) is shown in figure 14.

That is why the idea of the undercutting method was developed in the Department of Mining, Dressing and Transport Machines, AGH University of Science and Technology, in order to design an innovative construction of mining head, equipped with mini asymmetrical disk tools [1, 5].

At first an attempt was undertaken to work out a solution of a mining unit equipped with mini disk tools, mounted directly on the head body. Due to high load of the disk tools and low effectiveness of mining, this solution was given up. As a result of thorough analysis of the world technique condition, and results of own tests of rocks mining with asymmetrical disk tools (of diameter up to 160 mm), the Department started tests to devise a new conception of a mining head.

Figure 14. Visualization of a miner MTM 6 made by Aker Wirth with mining head equipped with 6 disc arms and MMM Modular Mining Machine made by Atlas Copco with a disc milling head, designed for Rio Tinto Company [2, 10].

In this design motion of tools will be forced, and will cause mining of a rock body with tools along complex movement trajectory. It allows crossing of mining lines of individual disk tools, and facilitates mining compact rocks through breaking off rock furrows. It should lower energy consumption of the process [6, 11]. Disk tools were mounted on separate plates that could rotate on the mining head body, and are propelled independently from it. The works were performed in cooperation with the REMAG Company. It was planned to work out and adapt the new head solution for a medium type roadheader KR150. The view of spatial model and ready mining head mounted on the KR150 roadheader is shown in figure 15.

The preliminary tests were performed on artificial large size concrete block of uniaxial compressive strength up to 80 MPa prepared at REMAG Company. The best effects of the work – large graining of the output, low engines load, and limited vibrations were obtained for the head body rotations at counter clockwise direction of value 20 rpm, and the plates rotations at clockwise direction of value 60 rpm. The change of direction of the head body or plates with disk tools rotation onto the opposite had a highly negative influence on the engines load and disk tools, and the plates wear.

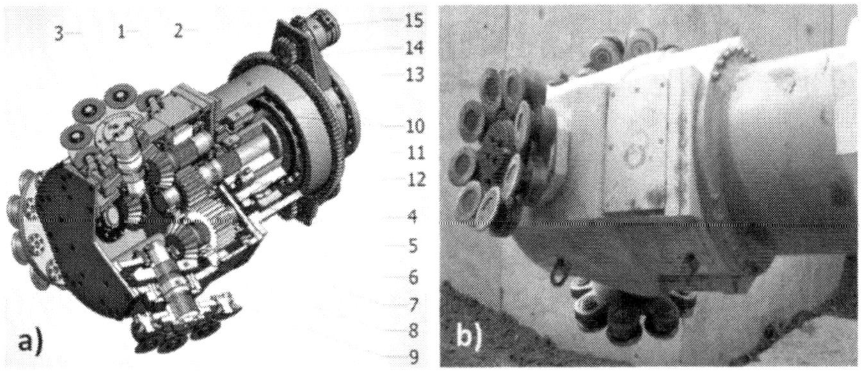

Figure 15. The new mining head solution with disc tool of complex trajectory: a) – spatial model, b) - view of KR150 roadheader mining head ready before field tests [11]. 1 – main gear, 2 – auxiliary gear, 3 – disc plate, 4 – input shaft, 5 – contral gear wheel, 6 – orbital gear wheel, 7 – pinion, 8 – face gear, 9 – output shaft, 10 – support, 11 – connector, 12,13 – bearings, 14 – body gear, 15 – hydraulic engine.

In order to check the influence of mounting and the deflection angle of the rotational axis of the disk tool, new exchangeable holders of these tools were made which allow disk tools mounting at an angle of -5 °, 45 ° and 90 ° relative to the surface of the plate. The view of the plates with disk tools with these holders during field tests are shown in figure 16. Only the disk tools mounted at an angle of -5 ° allowed for proper work, without significant wear [12].

Figure 16. A view of the plates with disk tools mounted in the holders at an angle of: a) 45°, b) 90° and c) -5 ° relative to the surface of the plate [12].

During the last phase of the field tests, the mining head was mounted on a larger roadheader FR250 - manufactured by FAMUR company, with more weight - over 65 tones and equipped with a larger power electric motor - 250 kW. This allowed for better, more stable operation of the mining head and the possibility of mining at greater depths. The view of this roadheader and mining head before the trials is shown in figure 17[13].

Figure 17. A view of FR250 roadheader with new mining head with disk tools before the field tests [13].

During the tests, using disk tools made of different materials and by various methods, the efficiency of mining was checked. Disk tools had the same diameter - 170 mm and wedge angle of 45°. They were made of tool steel, Hadfield or low-alloyed (GS42CrMo4) cast steel and ADI spheroidal cast iron. The tools manufactured from tool steel type NZ3 were full-hardened, to a hardness of over 55 HRC. The tools manufactured from tool steel type 36HNM were surface- hardened and tempered. Tools made of cast steel and cast iron were machined after casting. After this mechanical treatment, it was noticed that especially tools made of cast steel had numerous castings defects - blisters and pores. These tools have been disqualified.

During concrete block mining with larger cutting depth, the best results were obtained for tools made from tool steel type 36HNM. After one hour operation there were not significant traces of wear observed. In the case of disk tolls made from tool steel NZ3, already after very short work numerous edge breaks as well as cracking and breakage of the part of tools were noticed.

However, in the case of disk tools made of ADI cast iron, no breakage of the tool edge but its systematic abrasive wear was observed. The outside diameter of cast iron tools after about 30 minutes of work was decreased to value of about 135-140 mm. This diameter value did not allow further work. The wear view of tools made from tool steel NZ3 and ADI type cast iron is shown in figure 18. The reason of quick tools wear was probably to large cutting depth and movement speed of mining head. Combination of both of these parameters caused a heavy load on the tools [13].

Figure 18. A view of disk tool wear: A) made from steel type NZ3, B) made from ADI cast iron [13].

4. Summary

The mining of compact rocks with roadheaders with milling bodies seems to be currently the most popular method. As of today, attempts to use mini disk tools (up to 180 mm in diameter) on the roadheaders mining heads offer promising results. However, the biggest drawback in their case is durability. The implementation of these tools, for example, using the forging method and the appropriate heat treatment can give good results.

However, symmetric disk tools with a diameter of 400 - 500 mm, working with static crumpling, mounted on the mining heads of TBM machine, have very high durability but in this case very high value of pressure force require huge dimensions, weight and costs of the whole machine. This makes them economically viable in the case of large (over a few kilometers) runnout of excavations. The undercutting method using non-symmetrical disk tools with a diameter of about 400 mm, mounted on swing arms, offer also promising results. The new Aker Wirth machine are able to mine excavations in compact rocks with reduced energy consumption of the mining process and high granulation of the output.

References

[1] Kotwica K and Klich A 2011 Maszyny i urządzenia do drążenia wyrobisk korytarzowych i tunelowych *Instytut Techniki Górniczej KOMAG Gliwice*
[2] Brochures and information materials of Aker Wirth
[3] Brochures and information materials of Herrenknecht
[4] Brochures and information materials of Sandvik
[5] Klich A at all 1998 Niekonwencjonalne techniki urabiania skał. *Wydawnictwo Śląsk, Katowice*
[6] Kotwica K at all 2016 Wybrane problemy urabiania, transportu i przeróbki skał trudnourabialnych, Part 1 *Wydawnictwa AGH*
[7] Brochures and information materials of Mannesmann Demag
[8] Brochures and information materials of Robbins
[9] Weber W 1994 Drążenie chodników o różnych przekrojach przy pomocy techniki tylnego wycinania - Maszyna firmy Wirth-HDRK „Continuos Mining Machine" Sympozjum TMB Herne nt. Drążenie chodników w górnictwie węgla kamiennego. Siemianowice Śl
[10] Brochures and information materials of Atlas Copco
[11] Kotwica K 2018 Atypical and innovative tool,, holder and mining head designer for roadheaders used to tunnel and gallery drilling in hard rock T*unneling and Underground Space Technology* **8**, pp.493-503
[12] Kotwica K at all 2017 Wybrane problemy urabiania, transportu i przeróbki skał trudnourabialnych, Part 2 *Wydawnictwa AGH*
[13] Kotwica K at all 2018 Wybrane problemy urabiania, transportu i przeróbki skał trudnourabialnych, Part 3 *Wydawnictwa AGH*